T0370372

Renewable Energy

Peter Yang

Renewable Energy

Challenges and Solutions

Peter Yang
Case Western Reserve University
Cleveland, OH, USA

ISBN 978-3-031-49127-6 ISBN 978-3-031-49125-2 (eBook)
https://doi.org/10.1007/978-3-031-49125-2

This Springer imprint is published by the registered company Springer Nature Switzerland AG
The registered company address is: Gewerbestrasse 11, 6330 Cham, Switzerland

Paper in this product is recyclable.

This book is dedicated to my wife Yuezeng Yang who inspired me to take on this project through patience and questions.

About This Book

This book informs the college students and young professionals who have no renewable energy background on energy revolution and renewable energy technology.

The humanity is facing urgent needs for energy revolution: We must act on the Paris Agreement to limit global warming to well below 2 °C, preferably to 1.5 °C to avoid the irreversible environmental tipping point. To reach this objective, nations must replace the fossil fuels with 100% renewable energy as our major defense mechanism against global warming and climate change. We are running with the time against the calamitous global warming and climate change.

In this unprecedented mission, the world is dealing with significant challenges in the deployment of renewable energy. As intermittent and variable energy sources, renewable energy power generation and supply tend to be less dependable and stable. To make renewable energy generation and consumption much more efficient, dependable, stable, and cheaper, much more coordinated technological, political, socioeconomic innovations are necessary at international, national, and local levels.

To accomplish this mission, the United Nations' Greening Education Partnership program called for "Getting every learner climate-ready" through four pillars of transformative education of greening schools, curriculum, teacher training and education systems' capacities, and communities. This book aims to fill this gap in our urgent climate action. Each chapter is writing for the target academic knowledge level of college students and young professionals without background in renewable energy technologies. At the end of each chapter, there are activities related to the benefits, challenges, and solutions the renewable energy technology featured in the chapter. The answer key is presented at the end of the book.

Contents

About the Author

Peter Yang is an accomplished author, editor, researcher, and teacher in Sustainable Development, Renewable Energy, and German Studies. His current research focuses on climate change and climate action, and more specifically, the fossil fuel-based economic causes of climate change in the major economies and their actions to mitigate CO_2 emissions, including the deployment of renewable energy and energy efficiency technologies. Research projects he has completed include environmental and ecological impact of carbon-based energy production and consumption in major economics; investment, installation, and consumption of renewable energy technologies; and renewable energy promotion policies and regulations, such as sustainable development goals, renewable energy targets, carbon reduction targets, feed-in tariffs, fuel taxes, and carbon taxes. These projects resulted in three books, *Cases on Renewable Energy and Sustainable Development* (IGI-Global, 2019), *Rolling Back the Tide of Climate Change: Renewable Solutions and Policy Instruments in the U.S.A. and China* (Green Economics, 2015), and *Renewables Are Getting Cheaper* (Green Economics, 2016), and many refereed journal papers, book chapters, book reviews, and conference papers. His current energy-related research interests include Sustainable Development Goals, challenges and solutions of renewable energy technologies, grid integration, and energy storage; energy efficiency in transportation and buildings; R&D of renewable energy technologies; as well as teaching, training, and public education of renewable energy transformation.

Abbreviations

AC	Alternating current; air cooling or conditioning system
ACS	Absorption cooling system
AGS	Advanced geothermal system
AI	Artificial intelligence
ARENA	Australian Renewable Energy Agency
ASCE	American Society of Civil Engineers
a-Si	Amorphous silicon
BEST	Better Energy Storage Technology
BIPV	Building-integrated photovoltaics
BYD	"Build Your Dreams," EV manufacturer
CAES	Compressed-air energy storage
CAGR	Compounded annual growth rate
CAPEX	Capital expenditure
CdTe	Cadmium telluride
CEFC	Clean Energy Finance Corporation
Cefic	European Chemical Industry Council
CIGS	Copper indium gallium selenide
CO_2	Carbon dioxide
COD	Chemical oxygen demand
CPV	Concentrated photovoltaic
CSP	Concentrated solar power generation
ct	Cent
DC	Direct current
DOE	Department of Energy
e.g.	For example
EGS	Enhanced or engineered geothermal systems
E*n*	*n*% ethanol (e.g., E10: 10% ethanol)
EPA	U.S. Environmental Protection Agency
ESGC	Energy Storage Grand Challenges
ETC	Evacuated tube collector
EU	European Union
EuCIA	European Composites Industry Association
EV	Electric vehicle
F	Fahrenheit; farads

FAME	Faster Adoption and Manufacturing of Hybrid and Electric Vehicles
Fe	Iron
FIT	Feed-in tariff
FT	Fischer–Tropsch
g	Gram
GDP	Gross domestic product
GDU	Geothermal direct use
GFPC	Glazed flat-plate collector
GW	Gigawatt, thousand MW
GWh	Gigawatt hour
HeatORC	Heat organic Rankine cycle
Hg	Mercury
IEA	International Energy Agency
IFPSH	International Forum on Pumped Storage Hydropower
IISD	International Institute for Sustainable Development
IRENA	International Renewable Energy Agency
J	Joule
kg	Kilogram
kW	Kilowatt (thousand watts), unit of power
kWh	Kilowatt hour, unit of energy
LCOE	Levelized cost of energy
Li-ion	Lithium ion
MSW	Municipal solid waste
MW	Megawatt, thousand kilowatts
MWh	Megawatt hour, thousand kWh
NEMMP	National Electric Mobility Mission Plan
NGK	Sodium sulfur
NiMH	Nickel metal hydride
NO_x	Nitrogen oxides
NREL	National Renewable Energy Laboratory
NSF	National Science Foundation
O&M	Operation and management
ORC	Organic Rankine cycle
PHS	Pumped hydropower storage
PPA	Power purchase agreements
PV	Photovoltaics
PVT	Photovoltaics thermal collector
R&D	Research and development
RE	Renewable energy
REN21	Renewable Energy Network for the 21st Century
s	Second
SAF	Sustainable aviation fuel
SDGs	Sustainable Development Goals
SDH	Solar district heating
SDHC	Solar district heating and cooling

SEI	Solid-electrolyte interphase
SEIE	Solar Energy Industries Association
SHC	Solar Heating and Cooling Program (IEA)
SHW	Solar hot water
SMES	Superconducting magnetic energy storage
SO_2	Sulfur dioxide
SRAM	Static random-access memory backup
TES	Thermal energy storage
TW	Terawatt, thousand million (billion) kW, thousand GW
TWh	Terawatt hour, thousand million (billion) kWh
U.S.	United States
UHVAC	Ultrahigh-voltage alternating current
UHVDC	Ultrahigh-voltage direct current
UPS	Uninterruptible power supply
VAT	Value-added tax
VRE	Variable renewable energy
W	Watt
WETO	Wind Energy Technologies Office
Wh	Watt hour

List of Figures

Solar Power 1

Introduction

Science and Technology

1. Solar Photovoltaic Power

Solar photovoltaic (PV) power is a modern renewable energy technology consisting of solar PV cells, modules (panels), and arrays (systems, installations). This modular technology is used to generate electric power by converting sunlight (solar energy) into electricity (Fig. 1.1).

Solar cells are the basic building blocks of photovoltaic (PV) power generation systems. These cells are made of semiconductor materials, typically silicon, which converts sunlight into electrical power.

The conversion of sunlight into power in a solar cell occurs through the PV effect. This effect occurs when photons, or particles of light, strike the surface of the solar cell and are absorbed by the semiconductor material. When this happens, electrons in the material are excited, which allows them to move freely within the material.

To generate power in the solar cell, the excited electrons need to be collected and harnessed. To do this, the solar cell is constructed with two layers of semiconductor material, one layer of which is doped with boron to create a region with an excess of electrons, known as the n-type layer. The other layer is doped with phosphorus to create a region with a deficiency of electrons, known as the p-type layer (Fig. 1.2).

When sunlight is absorbed by the solar cell, it creates an electric field between the n-type and p-type layers. This electric field causes the excited electrons to move from the n-type layer toward the p-type layer, creating a flow of electrical current. This flow of current can be harnessed and used to power electrical devices.

P. Yang, *Renewable Energy*, https://doi.org/10.1007/978-3-031-49125-2_1

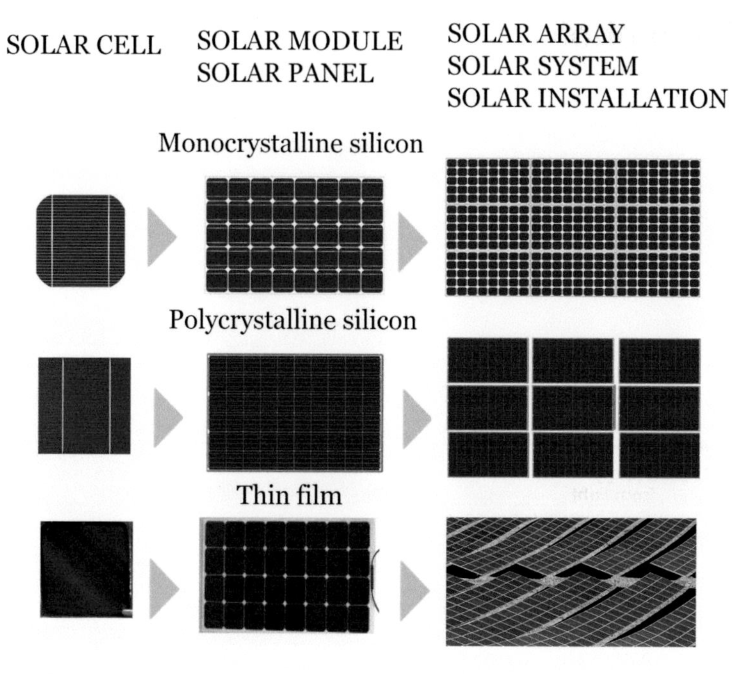

Fig. 1.1 Solar cell, module, array

2. Solar Installation

To generate useful amounts of power, solar cells are typically arranged in arrays and connected to form PV modules or panels. These panels can be installed on rooftops, on the ground, or on other structures to generate power from solar energy. The power produced by the panels can be used to power homes, businesses, and other electrical devices, or when there is surplus power supply, it can be fed into the grid or stored in batteries for later use.

Benefits

1. Abundant and Free Energy Source

Solar energy is a powerful, abundant, and inexhaustible renewable energy source. In just one and a half hours, the sun irradiates the Earth's surface with energy that exceeds the world's total energy consumption for a whole year. Since large amounts of energy come continuously from solar rays, there is no concern about the depletion of the energy source. Unlike fossil fuels such as coal, oil, and natural gas, solar energy is freely available, and solar PV power generation does not have an additional energy source cost. Considering the finite reserves of other energy

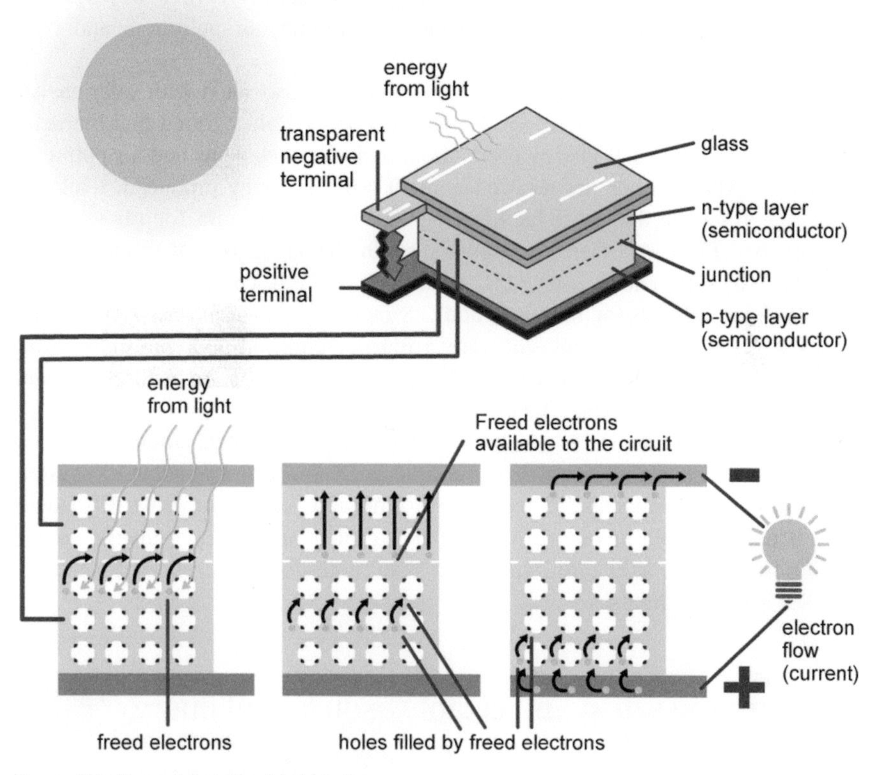

Fig. 1.2 How the PV cell generates power

sources, especially fossil fuels such as oil, gas, and coal, the large amount of solar energy for PV power generation is the single most important natural endowment for the energy security of humanity.

2. Environment and Health Benefits

Solar energy, including solar power and solar heat, is the safest and cleanest energy source. Solar power is an emissions-free source of energy. Its production and consumption do not produce any greenhouse gas emissions. Although the manufacture and installation of solar power generating systems, such as solar cells, have environmental impacts, these impacts are minimal compared to fossil fuel-based power generation. The emissions-free nature of solar power is because it relies on the sun's energy to generate power rather than burning fossil fuels.

In contrast, the use of fossil fuels such as coal, oil, and natural gas for power generation emits large amounts of carbon dioxide (CO_2) and other harmful pollutants such as sulfur dioxide (SO_2), nitrogen oxides (NOx), and particulate matter. These emissions contribute to global warming, air pollution, and other environmental and health problems.

Various studies have confirmed the low death rate associated with solar energy. For example, a study published in the journal Energy Policy found that the death rate associated with solar energy production, including accidents and air pollution, was only 0.004 fatalities per terawatt-hour (TWh) of energy produced, while the death rate for coal was 24.6 fatalities per TWh, and the rate for oil was 18.4 fatalities per TWh. Another study published in the same journal found that the death rate for solar energy was 0.002 fatalities per TWh, compared with 24.6 for coal, 18.4 for oil, 4.6 for biomass, and 2.8 deaths for natural gas. At the same time, the greenhouse gas emissions are 5 ton per gigawatt-hour power for solar PV power versus 820 ton for coal, 720 ton for oil, 490 ton for natural gas, and 78–230 ton for biomass (Fig. 1.3).

Therefore, the transition to solar PV power offsets enormous amounts of emissions of carbon and pollutants. For example, an average 4 kW solar PV system could offset approximately 1.5 ton of CO_2 emissions each year. Solar PV power plays a decisive role in combating global warming and climate change caused by CO_2 emissions, acid rain, smog, haze, air pollution, and related respiratory illnesses and lung diseases.

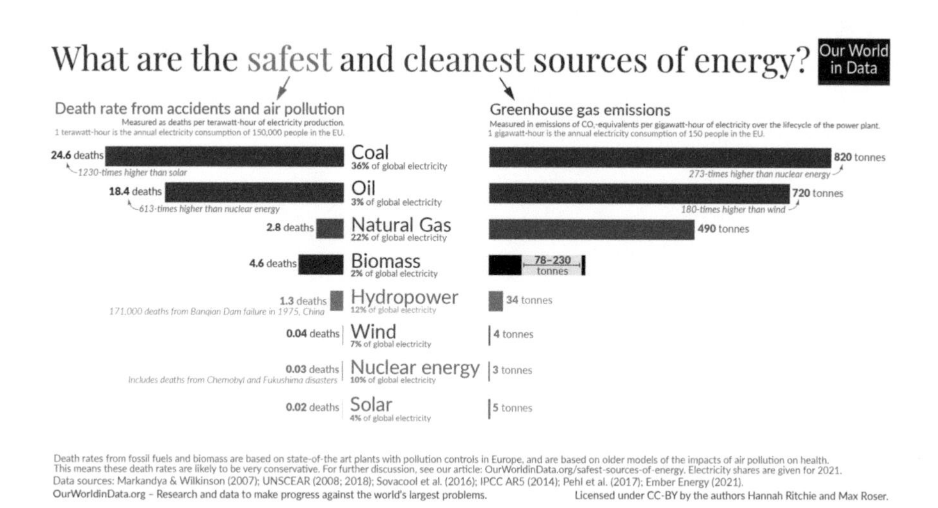

Fig. 1.3 Social and environmental impact of power generation

3. Flexibility and Scalability

The modular implementation of solar PV technology makes the installation of PV power generation systems flexible and scalable. PV installations can take any area exposed to solar radiation. This means that PV systems can be ground-mounted, rooftop-mounted, floating, or wall-mounted. Most solar panels are mounted on a rack in a fixed position and angle facing the sun. However, to optimize power output, solar modules can also be mounted on one or two raised poles, which allows them to track the sun following the sun's movement across the sky. Solar PV installations are also fully scalable because the size of a PV system can be increased whenever the demand for power increases.

Deployment

A. Solar Power Deployment

1. Early Research and Development

Solar cell technology, the core of solar PV power generation, has a long history. French physicist Alexandre-Edmond Becquerel was the first scientist to discover the PV effect in 1839 when the exposure of metal electrodes to light generated small electric currents. In April 1954, researchers at Bell Laboratories introduced the first practical silicon solar cell.

Early solar cells were made from single-crystal silicon in the 1970s. These monocrystalline solar cells had efficiencies of only approximately 4%, and they were so expensive that they could only be used in niche applications such as space exploration. In the 1980s, polycrystalline silicon solar cells were introduced, which had efficiencies three times those of early monocrystalline solar cells. However, their production cost remained too high to allow them to be widely adopted. In the 1990s, the focus was on p-type and n-type mono- and polycrystalline solar cells, which brought considerable advancements to solar cell technology. These cells had efficiencies of approximately 15%, which made them more cost-effective and suitable for commercial applications.

2. Recent Surge

Solar PV power generation is the fastest-growing energy industry in the world, and it has achieved a great development level driven by generous government financial support in many countries. From 2013 to 2022, the cumulative installed capacity of solar PV generation grew from 137 to 1047 GW, accounting for an annual average growth rate of 26% in the last decade (Fig. 1.4).

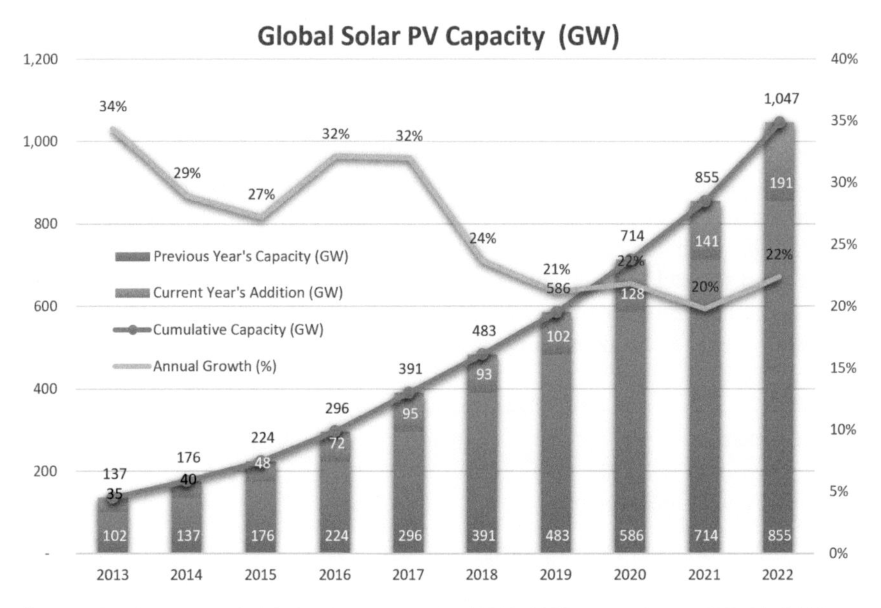

Fig. 1.4 Rapid growth of global solar PV capacity (2013–2022). *Data Source* IRENA (2023)

The deadlines of the government support policy in countries such as China and the U.S. led to a PV deployment boom in 2021. In China, annual solar PV capacity additions have rebounded since 2020 after a slowdown in the 2018–2019 COVID-19 pandemic. The country added 86 GW of solar PV capacity and reached 392 GW, which accounted for an annual growth rate of 28% in 2022.

The top four solar PV power generating countries—China, the U.S., Japan, Germany, India, Australia, and Italy—maintained their relative rankings. However, India succeeded in achieving an annual growth rate of 27% and reduced its gap with Germany. Brazil was able to increase its PV capacity by 70% and take eighth place. The Netherlands also enjoyed an excellent 51% increase in the expansion of its solar PV capacity.

The solar PV industry in China experienced a rapid expansion from 3% in 2010 to 37% in 2022. The U.S. and Japan had relatively stable shares, and Germany had a declining share in global solar PV deployment in recent years. The installed capacity (in GW and share) of the solar PV installations of the world's top four and other countries during 2010 and 2022 is shown in Fig. 1.5.

B. Advancements and Impacts

1. Economies of Scale

Most advancements in solar cell technology have taken place in the past two decades as manufacturers have focused on improving the efficiency of these cells. The most efficient p-type and n-type mono- and polycrystalline solar cells

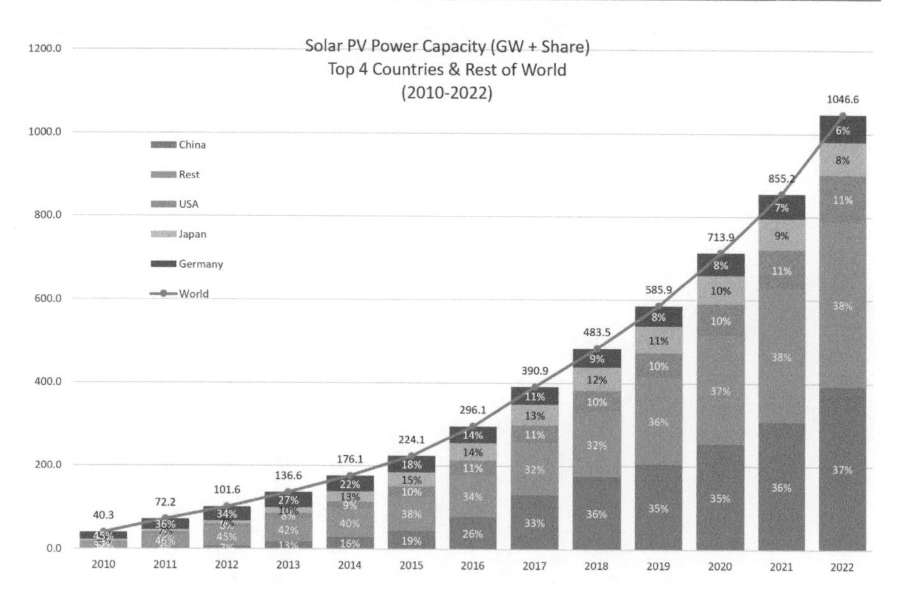

Fig. 1.5 Solar PV capacity of the top 4 and other countries, 2010–2022. *Data Source* IRENA (2023)

reached efficiencies of approximately 22% and 24%, respectively, by 2010 and approximately 26% and 27%, respectively, by 2020. These advancements were accompanied by a significant reduction in the cost of solar cell production, with the cost per watt of solar panels falling from over $70 in the early 2000s to under $0.80 in 2020.

Today, monocrystalline panels are the most popular solar panels on the market. They offer the highest efficiency rate, at an average of 24%. However, with an average cost ranging from $1 to $1.50 per watt, monocrystalline panels also cost much more than polycrystalline panels. With an average cost for polycrystalline solar panels of $0.90 to $1 per watt, polycrystalline panels are more cost-effective than monocrystalline panels. However, their efficiency rate is also lower than that of monocrystalline panels, at an average of 20%. Thin film solar panels are increasingly being installed on the rooftops of residential, commercial, and industrial buildings. In addition, they are also commonly used in large-scale solar farms or utility-scale installations. They tend to cost approximately $1 to $1.50 per watt.

When comparing the deployment of p-type and n-type mono- and polycrystalline solar cells, it is recognizable that most research and development efforts in the past 20 years have focused on p-type cells because these cells were chosen for space exploration applications due to their better performance in resisting radiation. However, in recent years, the advantages of n-type cells in terms of performance and durability have become more apparent, leading to increased interest in this technology, and the focus has notably moved from p-type cells to n-type cells.

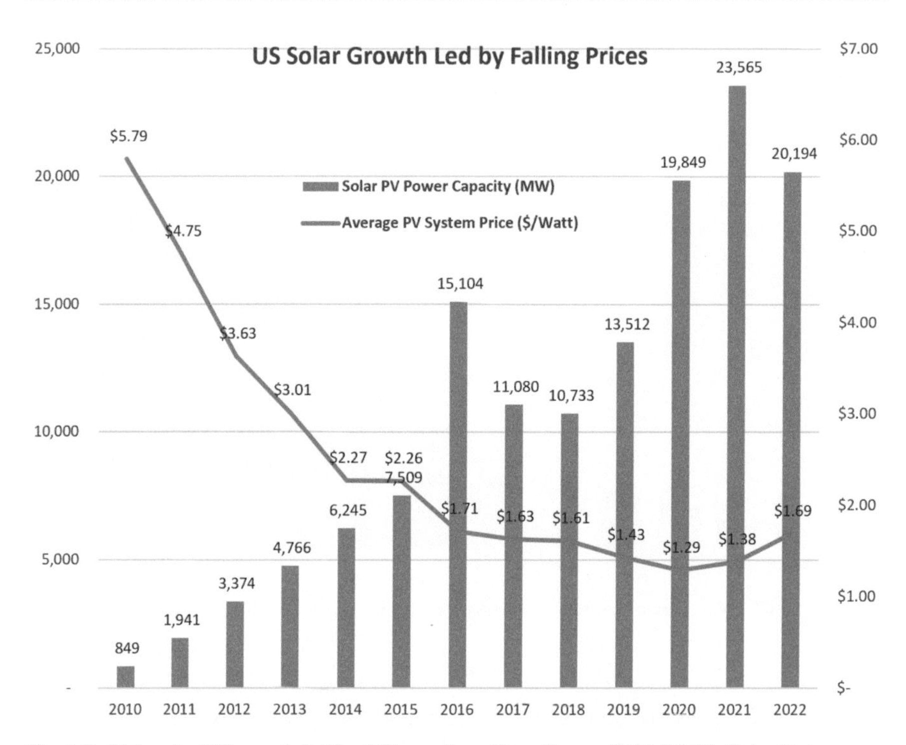

Fig. 1.6 U.S. solar PV growth led by falling prices. *Data Source* SEIE (2023). Solar Data

Solar PV panels were extremely costly decades ago. However, PV module market data show that in recent years, global PV module production increased by six times, and the solar PV module price dropped seven times during 2010–2022. Figure 1.6 shows the PV power capacity increase and the PV power price decrease in those years.

The rise of China's solar PV manufacturing and the massive investment in solar PV installations combined allowed its PV power generation to experience the fastest price drop among all renewable energy technologies in the country and the rest of the world. In the last decade, the cost of PV power generation worldwide dropped by 90%. This drastic market dynamics is an excellent example of economic theories of *economies of scale* and *learning curves*. Box 1 shows these economic theories on solar PV growth and price drops. These theories can also be adaptively used to explain the relationship between the growth of other renewable energy technologies and price drops.

Box 1 Economic Theories on Solar PV Growth and Price Drop
The theory of economies of scale indicates that the market price of a product drops with an increased scale of its installation, production, and marketing. The theory of the learning curve or experience curve was developed by Richard Swanson, founder of high-efficiency solar panel manufacturer SunPower. It states that several improvements, including learning, experience, and technological innovation, will improve economic efficiency, market expansion, and price reduction, which causes a relationship between every doubling in global solar PV capacity and a 20% price reduction of solar PV modules.

In summary, the mass production of PV panels has led to economies of scale, reducing the cost per unit. As demand for solar energy has increased worldwide, the production volumes have also grown, enabling manufacturers to achieve cost efficiencies through large-scale production.

2. Sharpest Price Drop Among Power Generation Technologies

The dramatic price drop in the last ten years allowed solar PV power to demonstrate its price advantage over other energy sources. As of the end of 2021, the *levelized cost of energy* (LCOE, total investment cost divided by the total power generated over the equipment's lifetime) of utility-scale solar PV power per megawatt-hour was $36, which was only one tenth of the LCOE in 2009. In addition, utility-scale solar generation became the cheapest among the levelized costs of all energy sources, including fossil fuels (see Fig. 1.7).

When compared with other power generation technologies, including other renewable energy technologies, fossil fuels, and nuclear power, PV has experienced a more rapid decline in costs. The sharp difference in price drop between PV power generation and other technologies can be attributed to technological advancements, declining module prices, economies of scale, policy support, and industry experience. These factors have collectively driven the rapid cost reduction of PV, making it a highly competitive and attractive option for clean and sustainable energy generation.

Advancements in PV technology have played a crucial role in reducing the cost of solar power generation. Improved manufacturing processes, increased efficiency of solar panels, and economies of scale in production have all contributed to the cost reduction. As technology continues to improve, the cost of PV power generation is expected to decrease further.

The consistent cost reduction of PV modules, which are essential components of solar systems, is mainly due to increased competition among manufacturers, improved production methods, and advancements in material technologies. As a result, the overall cost of PV power generation has significantly dropped.

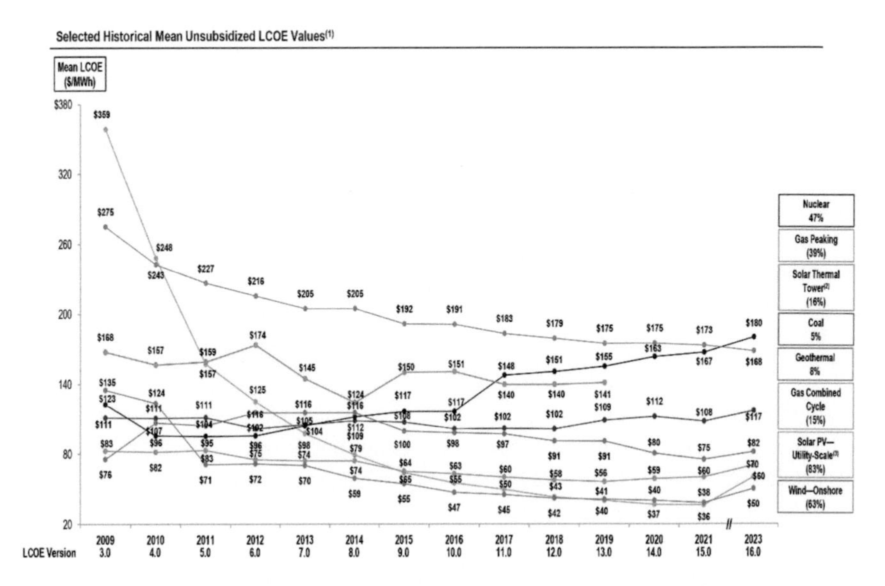

Fig. 1.7 Cost trend of power generation by energy. *Image Credit* Lazard, 2023

Most importantly, the policies and incentives many governments have implemented to promote solar PV power, including feed-in tariffs, tax credits, grants, and subsidies, have encouraged the adoption of PV systems and boosted the demand for PV installations, leading to increased production and lower costs.

As the PV industry has matured, manufacturers have gained experience in producing solar panels, installing systems, and managing solar projects. This increased industry experience has led to improved efficiency, reduced costs, and streamlined processes throughout the value chain.

3. Government Support

Counterintuitively, the initial solar PV takeoff in the world's least sunny region of Germany occurred more than twenty years ago. It was driven by the German government's pioneering renewable energy promotion legislation and related innovative solar power generation incentives in the form of the feed-in tariff (FIT, an overmarket power price paid by the utilities to the solar power generator when the power is sold to the grid). The German FITs) had many innovative and effective features to promote the deployment of every renewable energy technology by every social and economic sector.

First, German FITs were designed according to the technical maturity level of the individual renewable energy technologies. The FITs were much higher for solar PV power generation than for wind power or hydropower to help investors afford the much higher installation cost of solar PV power. In addition, the FITs for residents' installations were higher than those for utility-scale installations, which helped create approximately 1.9 million solar PV prosumers[1] in Germany as of the end of 2021. The pioneering German FITs also had gradually reducing rates over a 20-year period, helping Germany win and maintain the world's top solar generator position for ten years until China overtook it in 2015.

In the aftermath of the U.S. financial crisis in 2008, the U.S. federal government provided *production tax credits* or *investment tax credits*[2] as part of economic recovery packages. The PTC/ITC provided a 30% tax credit for residential and commercial solar installations to help offset the high upfront costs associated with installing solar panels. Government support helped solar PV installations experience exponential expansion and price drops in the U.S. between 2009 and 2016.

Massive renewable energy incentives in the form of FITs also helped many other countries witnessed exponential growth of solar power deployment. The Chinese government has provided various incentives for solar PV power generation, including subsidies for renewable energy projects, FITs, and tax incentives.

These incentives helped China become the world's largest producer of solar energy in 2015, maintain this position ever since, and achieve grid parity. According to a recent IRENA study, the global average cost of solar PV electricity was $0.048/kWh in 2021. China had the lowest solar PV power cost at $0.034/kWh in 2021, which is lower than the average retail electricity price of $0.08/kWh in China in 2020.

The governments of many other countries, such as Japan, India, Italy, Spain, France, Australia, and Canada, have also implemented policies and incentives to support the growth of the solar industry.

4. Job Creation

The solar PV power sector is a great job creator. It typically generates more jobs per unit of power generated compared to other power generation sectors, such as power generation using fossil fuels. For example, solar PV creates approximately eleven jobs per MW of installed capacity compared with approximately 2.6 and 3.1 jobs per MW created by coal and gas-fired power plants, respectively, according to an IRENA study. This means that the solar PV sector creates approximately four times more jobs per unit of electricity generated than fossil fuels (Fig. 1.8).

According to a report by the IRENA, in 2020, the solar PV sector employed approximately 3.8 million people globally, representing a third of all renewable energy jobs worldwide. This is compared to approximately 1.1 million jobs in the wind energy sector and 2.5 million jobs in the bioenergy sector.

---

[1] A *prosumer* is defined as a consumer of power who is also producer of both the power for self-consumption and the excess power for sales to the grid to earn FITs, usually through a rooftop solar PV system.

[2] *Production tax credit* is referred to tax rebate on the solar power generator's tax liabilities at certain percent of the solar PV production cost. *Investment tax credit* is referred to tax rebate on the solar power investor's tax liabilities at certain percent of the solar PV system investment cost.

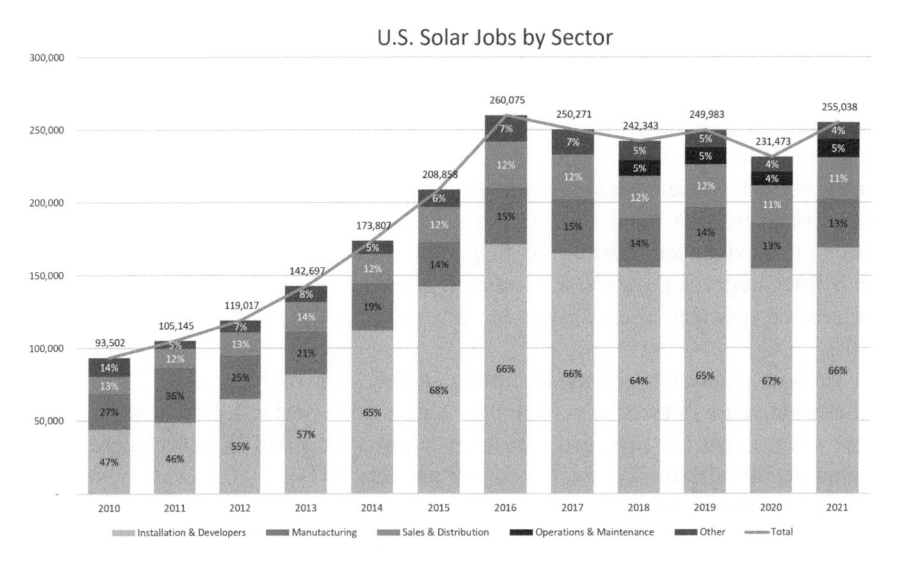

Fig. 1.8 U.S. solar jobs by category. *Data Source* SEIA (2022)

Challenges

A. Economic and Technological Challenges

1. Geographic Dislocation of PV Power Generation and Consumption

The distribution of solar resources is uneven. The global PV power potential map shows that large parts of Europe and Southeast China have far less PV power potential than the rest of the world (Fig. 1.9).

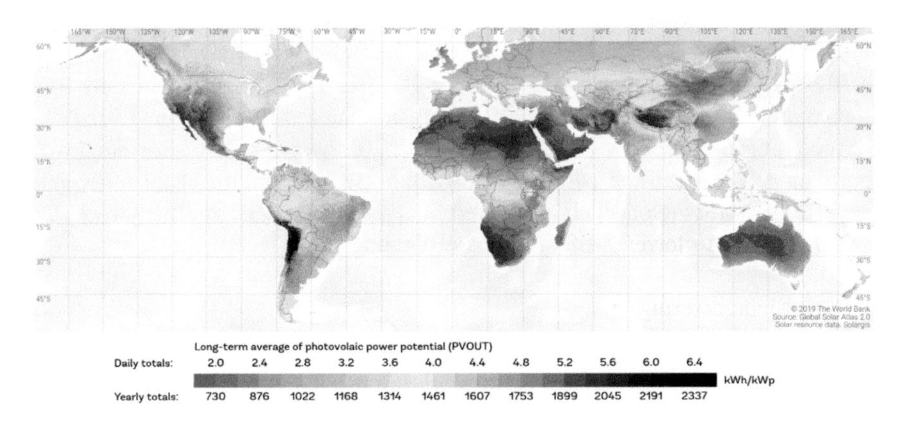

Fig. 1.9 Global average PV power potential

The uneven PV resource distribution can become an additional challenge to the further penetration of solar power when high PV resources are located in remote unpopulated deserts in Northwest China, while power-hungry cities are located in PV resource-poor Southeast China. This geographic mismatch of solar energy resources and economic centers causes a high curtailment rate of solar power in Northwest China because it cannot be consumed locally and because of the needs and costs for additional transmission grids to transmit solar power across China. High short-term and long-term variability in solar radiation and power output creates risks to the stability and reliability of the power grid. Interconnecting large amounts of variable solar power to the grid can also be a major challenge, especially when solar variable power generation is fully penetrated.

2. High Variability of Solar Radiation

In addition to geographic location, solar PV energy varies significantly depending on the time of day, weather conditions, and season. On sunny days, solar PV works best during the four or five hours around noon. Power output decreases rapidly in the early morning and late evening. At night, it does not work at all. On cloudy days or shorter winter days, the solar capacity factor decreases to 10–25% of that on sunny days. In contrast, when sunlight is intensive, massive power output can overwhelm a less powerful grid.

3. Insufficient Solar Cell Efficiency

Despite recent technological advances, the solar PV efficiency or conversion rate has ample room for improvement. The solar PV efficiency or conversion rate refers to the amount (as a percentage) of solar radiation that can be converted to power. Despite the incremental improvement, the current PV conversion rate remains insufficient.

The efficiency of *monocrystalline solar cells*, which are made of single-crystal silicon, ranges between 22 and 26%. This efficiency or conversion rate is typically higher than that of *polycrystalline solar cells*, which typically ranges between 17 and 22%.

Multijunction solar cells can achieve a much higher efficiency that reaches 30% for space applications and exceeds 40% for high-concentration applications.

Thin-film solar cells deliver even lower conversion rates, which range between 7 and 18%, depending on the specific material used. *Amorphous silicon (a-Si) solar cells* have an even lower conversion rate in the range of 6–10%.

4. High Manufacturing Cost

Monocrystalline and *polycrystalline solar cells* have relatively higher manufacturing costs due to the complex and energy-intensive processes and material requirements involved in producing crystalline cells using high-purity silicon. *Multijunction solar cells* have even higher manufacturing costs due to the complexity of the cell structure and specialized materials.

Thin-film and *amorphous silicon (a-Si) solar cells* have relatively lower manufacturing costs due to the simpler production process and use of fewer and cheaper materials, such as amorphous silicon (a-Si), cadmium telluride (CdTe), or copper indium gallium selenide (CIGS).

5. Low Temperature Tolerance

The heat of the solar panel caused by solar radiation hurts solar PV efficiency. Every 1 °C increase above 25 °C (77 °F) leads to a decrease in efficiency by 0.38%. Conversely, every 1 °C below 25 °C (77 °F) results in an increase in efficiency by 0.38%. This means that at the same sunlight, a higher temperature leads to a lower power output.

Monocrystalline and *polycrystalline solar cells* typically have lower temperature tolerance, which can cause a decrease in performance at high temperatures.

Thin-film and *amorphous silicon (a-Si) solar cells* generally have better temperature tolerance than crystalline cells, which allows them to perform relatively better in high-temperature environments.

Multijunction solar cells generally have better temperature tolerance than crystalline silicon-based solar cells. Their performance tends to be less affected by elevated temperatures, allowing them to maintain higher efficiency levels under hot conditions. These solar cells are typically composed of multiple semiconductor layers, each tuned to absorb different portions of the solar spectrum. This design allows them to achieve higher efficiency by capturing a broader range of light wavelengths. Multijunction solar cells are commonly used in *concentrated photovoltaic (CPV) systems*, where sunlight is concentrated onto a small area using lenses or mirrors. These systems involve higher operating temperatures due to the concentration of solar energy, and multijunction solar cells are specifically designed to handle the higher temperatures associated with concentrated sunlight, maintaining their efficiency under such conditions.

6. Limited Life Span

Monocrystalline and *polycrystalline solar cells* generally have longer lifespans, ranging from 25 to 30 years or more.

Thin-film and *amorphous silicon (a-Si) solar cells* have shorter lifespans than crystalline cells, typically approximately 20 to 25 years.

Multijunction solar cells may have a shorter lifespan due to material degradation of multiple layers of different semiconductor materials and higher heat stress from concentrated sunlight commonly used in *concentrated photovoltaic (CPV) systems*, which will be discussed in Chap. 2.

7. Urgent and Rising Green Power Demand

The global demand for solar PV power is expected to rise significantly soon because of several factors, such as the urgent need to keep the global temperature rise below 1.5 °C according to the Paris Agreement and energy sector integration with e-mobility, green hydrogen, and heat pumps. In Germany and other EU countries, there is also an urgent need for green energy to replace the current energy demand met by hard coal mines closed in 2018, nuclear power to be phased out by 2022, coal-fired power generation to be banned by 2030, and combustion engines to be banned by 2035.

At the same time, to meet the Net Zero requirements by 2050, the global cumulative solar PV capacity needs to be more than 14 TW, accounting for almost 20 times the current global cumulative solar PV capacity.

8. Market Challenges

The renewable energy policy switch from FIT to auction in Germany and China has exposed further solar PV energy expansion to greater market risks and challenges. The market challenges facing solar PV projects include technological lock-in,[3] power market segregation[4] designed for centralized power plants, market control by established fossil fuel power generators, and difficulties in overcoming their resistance, including slow interconnections and red tape of introducing solar PV systems, particularly for distributed rooftop PV power generation.

9. Inadequate Solar Power Penetration

Despite the rapid growth of global power generation, the share of the actual output of all solar PV systems in the global power output (called solar power penetration) remains low. The contributions of solar power to total power generation in 50% of the world's 10 top generating countries were below 5%, and those of the other 50% of countries were all below 11.4% in 2022 (Fig. 1.10). Worldwide, solar PV accounted for only 4.5% of global power generation in 2022. To achieve the world's Net Zero objectives, the share of global PV power in global power generation needs to be multiplied by 2050.

[3] A phenomenon where existing incumbent technologies dominate the market and create barriers for the adoption or development of potentially innovative alternatives or breakthroughs.

[4] Refers to the division or separation of a market into distinct segments based on certain characteristics or criteria.

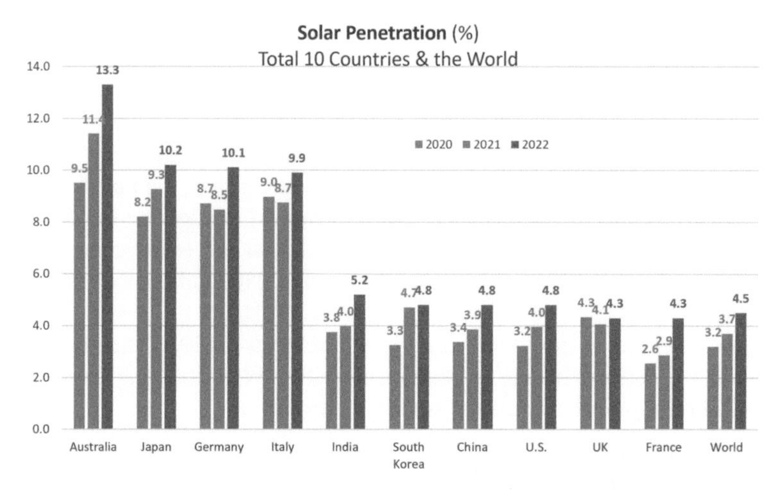

Fig. 1.10 Solar PV power penetration of selected countries. *Data Source* Our World in Data, 2023

10. High PV Investment and Soft Costs

Although sunlight comes to us as a free energy source, solar PV technology that turns it into power is costly. The high costs associated with PV installation include two parts—*hardware costs*[5] and *soft costs*.[6]

Over the last 20 years, solar PV costs have decreased by 90% thanks to government financial support. However, despite the dramatic drop in solar PV hardware costs (more than 60%) since 2010, the upfront solar PV costs, including soft costs, remain high. The high upfront cost and the perception of such a high price still discourage people from investing in solar PV technology.

The results of a recent survey show that more than 50% of the UK residents who did not invest in a solar PV system cited high investment costs as the main reason for their reluctance to invest. These findings have two implications. On the one hand, the upfront investment indeed remains a major concern for residential or business investors. On the other hand, it also indicates that the public is unaware of the dramatic falling solar PV price.

[5] *Hardware costs* include the costs of all the materials needed to construct the system: solar modules, inverter, racking, and electrical wiring. Solar modules or panels are expensive because they are made of large amounts of high-purity silicon, which requires a large amount of power.

[6] *Soft costs* include the cost of installation labor, the cost of all relevant permits, and all overhead costs including the marketing, sales and administrative costs associated with the system.

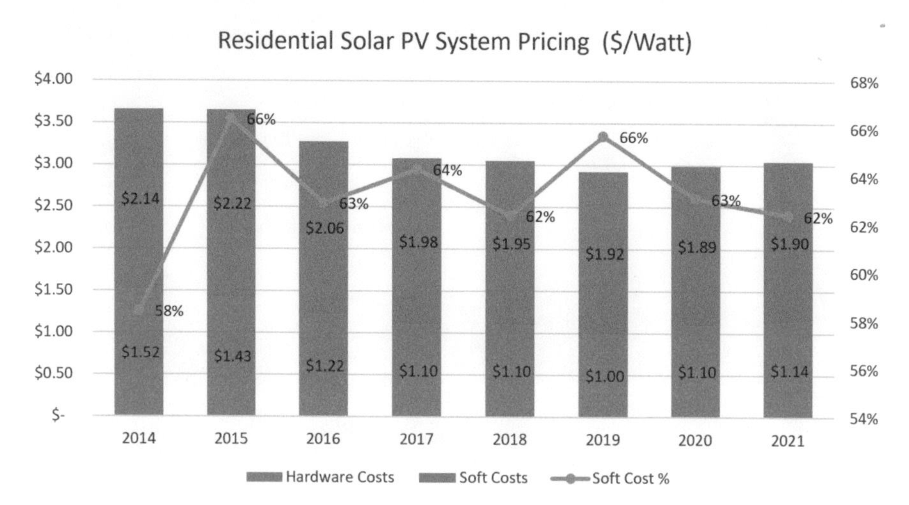

Fig. 1.11 U.S. PV system costs: hardware costs versus soft costs. *Data Source* SEIA (2022) Solar Industry Research Data

Different from the solar PV markets in developed countries, the utility-scale solar PV installations in China had a higher ratio of soft costs (70%) than distributed residential PV installations because utility-scale solar PV installations had an additional cost for land use. In the U.S., soft costs for residential PV installations made up 62% of the total solar PV installation in 2021 (Fig. 1.11), which was much higher than 15% in Germany and 25% in Australia.

11. Solar Curtailment

Solar curtailment takes place in solar abundant areas and is a power regulating tool used by the grid against the variable solar power when it is considered excessive and destructive for grid stability. Rising penetrations of variable solar PV power and the resulting increased variability in power systems are expected to increase solar curtailment—the forced reduction of solar power delivered due to oversupply or lack of system flexibility. Solar curtailment reduces the use of solar power, is considered a waste of renewable energy, and adversely impacts the further penetration of solar power.

Solar curtailment has been a major issue in China. Until recently, the curtailment rates in some remote solar power plants were as high as 50%. According to research, solar curtailment in China has a close relationship with the rapid development of the nation's utility-scale PV installations in remote areas, the high share of PV power generation in local power generation, long-distance grid transmission, inadequate grid transmission coordination, and high grid transmission costs (Fig. 1.12).

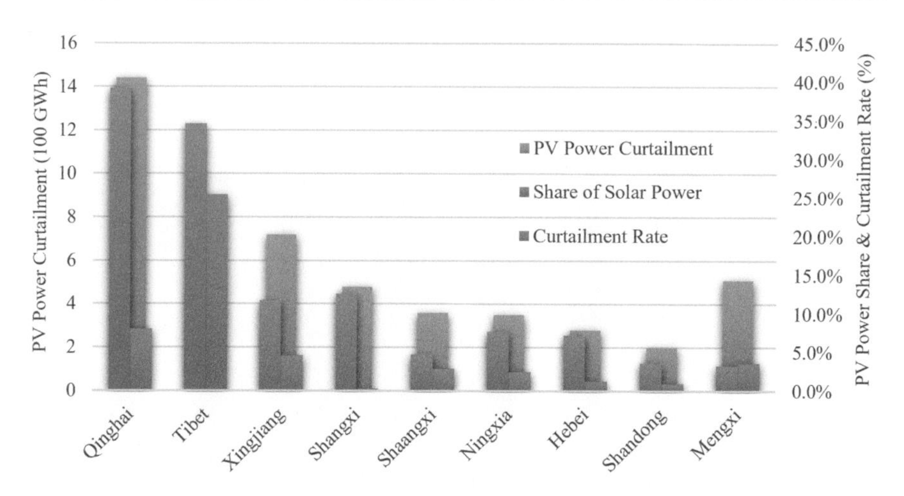

Fig. 1.12 PV share and curtailment of selected Chinese Provinces, 2020. *Data Source* China National New Energy Consumption Monitoring and Early Warning Center, 2021

Through several major steps to address the solar curtailment issue, the country was able to significantly reduce its solar curtailment rates. For example, its annual solar PV curtailment decreased from 10% in 2016 to 1.8% in 2022. However, with further solar penetration, China's curtailments are expected to increase.

In the U.S., solar curtailment has been less serious than in China, as the annual curtailment rate has been 4%. However, the curtailment rate for midday solar power can reach 10%, and the increased penetration of solar PV power tends to increase the solar curtailment rate. For example, according to California ISO renewable energy curtailment data, the solar power curtailment rate in California was 1.6% in 2018 but rose to 3.2% in 2019, 4.4% in 2020, 4.2% in 2021, and 6.3% in 2022.

12. Insufficient Storage Capacity

The FIT-supported solar energy installations in Germany and other countries mainly relied on grid integration. The current insufficient energy storage capacities around the world constitute a major challenge to the further expansion of increasingly dominant intermittent solar power. Pumped hydropower storage, which is the most common, affordable, and fast-expanding form of grid power storage, has geographical limitations.

Other forms of energy storage are available but still too expensive, and their capacity is too limited to serve as grid energy storage. Energy storage capacity building is still at the initial stage. The lack of energy storage causes an ironic contradiction between the decreasing solar energy costs and the increasing higher power bills for consumers and small businesses. There is an urgent need for research into completely new types of energy storage devices with potentially novel materials and physical and chemical characteristics.

13. Utility-Scale Solar Power Challenges

In contrast to Germany and the U.S., which have a larger share of distributed solar PV power generation than that of utility-scale solar PV power generation, utility-scale solar PV power generation in China has a significantly larger share than that of distributed solar PV power generation. In addition, in contrast to these developed countries, which have cheaper utility-scale solar PV power prices than distributed ones, the overall cost of utility-scale solar PV in China is higher than that of distributed solar PV.

The reversed ratios in China were caused by two major factors. First, a utility-scale solar PV power plant needs a much larger area of land than distributed solar PV installations. The additional costs of land use were, especially in economically developed areas, not only difficult to reduce but were increasing every year. Second, when the utility-scale solar power plant was installed in Northwest deserts, there was a thousand-mile-long distance between the plant and its final power users in Southeast China, and long-distance power transmission was associated with additional costs and power loss.

14. Inadequate Transmission Grid

Transporting utility-scale solar PV power from remote solar power plants to urban load centers where most of the country's energy is consumed requires an adequate transmission grid. Intermittent solar resources and power output pose unique problems in transmission planning and efficient utilization of transmission infrastructure, resulting in higher transmission costs, increased congestion, and even power curtailment with a lack of adequate transmission capacity. Due to potential transmission constraints, solar project developers will need to evaluate the economic tradeoff of sitting where the resource is best versus sitting closer to loads where transmission constraints are less likely.

B. Political, Social, and Environmental Challenges

1. Insufficient Government Support

Considering the extremely high solar PV costs, support from the government was essential for existing deployment. Reducing or stopping this support is a challenge to solar penetration. Significant differences in countries' solar PV technology deployment levels were caused by different government and financial support levels. Inadequate financing support by governments and financial systems, a lack of financial options for solar PV projects, and insufficient stakeholder/community engagement in solar PV power initiatives were some of the main political challenges to solar technology dissemination.

In Germany, the rapid growth in distributed solar PV installation and generation driven by administratively set FIT incentive schemes was criticized for its lack of solar PV market efficiency and costs to governments and consumers. As a result,

Germany significantly reduced tariff levels, which in turn led to a slowdown in solar PV deployment. However, no auction was scheduled for utility-scale solar PV systems, nor was it clear if remuneration would be extended for distributed PV expansion beyond the 52 GW cumulative installed capacity cap set by the German government, which the German solar industry viewed as a greater threat than the COVID pandemic. Similar issues also exist in other countries.

With the increased price competitiveness of solar power technology and solar power curtailment, China also gradually moved away from government-set financial incentives and switched to market-determined promotion tools, such as auctions. Since 2020, the country has either significantly reduced or stopped government financial support for solar PV installations, and solar power developers have started to rely on their marketing efforts to make their investment decisions.

In the U.S., it is unclear whether the existing solar support for distributed solar will be approved by the U.S. Congress after its expiration. In addition, there is a lack of transparent policies and regulations supporting the penetration of solar technology, credible endorsements of PV to instill consumer confidence, siting, permitting utility PV programs and government tax credits, as well as information dissemination and consumer awareness about solar PV power.

2. Land Use

Solar radiation has a lower energy density than conventional energy sources and therefore needs large areas to receive solar energy. Land use is often an issue for different reasons. For geographically small and densely populated countries such as Germany and Japan, siting and permission for land use are difficult because of the limited available land and soaring land prices.

Box 2 Solar Land Use Versus Land Price Rise

Solar Land Use. Based on the current PV panel efficiency, an average utility-scale solar PV power plant requires approximately seven acres per megawatt (MW) of generating capacity; a 200 MW PV power plant would require approximately 1400 acres of land.

Land Price Rise in Germany. Because land area has become more and more scarce, land prices in Germany rose sharply. Agricultural land prices increased by 174% in Germany in 2018 from 2001, and rising land procurement costs for renewable power installations became one of the increasing obstacles to renewable energy growth.

Certainly, countries with a relatively large land area, such as China, might have more space for solar use, such as in remote deserts. However, the resulting geographic mismatch between solar power generation and consumption requires additional transmission equipment and cost. In the U.S., siting and permission of land use, as well as environmental review, takes more than 3–5 years.

3. Environmental Impact of the Solar Panel

Although solar energy sources and PV power generation are clean, the manufacturing and recycling of PV power equipment have significant environmental impacts. The materials used in solar panels, such as *nitrogen trifluoride*,[7] cadmium,[8] and lead, and the intensive use of currently carbon-intensive power for purifying silicon are environmentally impactful. The toxicity of materials in solar panels, such as the toxic substances used for processing PV materials, causes pollution and poisoning. The intensive use of mostly coal-fired power to purify silicon used in solar cells and transport and install solar power systems also increases CO_2 emissions and pollution. The disposal and recycling of end-of-life solar panel waste requires viable technologies. The use of large areas required by solar PV power generation is also a major environmental concern. The exclusive use of large areas of land for utility-scale solar power installations, for example, can cause habitat loss for both plant and animal species, limit agricultural and forest development, and lead to wild species extinction.

4. Inadequate Skilled Workforce and Training

The further penetration of solar PV power lacks the workforce with adequate scientific, technical, and manufacturing skills needed for solar PV installation, maintenance, and inspection. The current educational system failed to provide adequate education and training in solar technologies necessary for solar companies to easily recruit new hires and expand their solar installation business. These issues cause inadequacies in the installation, maintenance, and inspection services of solar PV systems and high soft costs.

Solutions

A. Economic Solutions

1. Further Reducing Solar PV Costs

A better understanding of solar costs. To address high upfront solar PV costs, we need a better understanding of the perceived high solar costs. First, we can compare the solar PV costs with fossil fuel and nuclear power costs, analyze the breakdowns of different types of solar PV costs, and manage future solar PV power costs.

[7] *Nitrogen trifluoride* finds increasing use in manufacturing PVs. However, it is an inorganic, colorless, nonflammable, and toxic gas with a slightly musty odor. It is also an extremely strong and long-lived greenhouse gas.

[8] Although the major use of cadmium in rechargeable nickel-cadmium batteries has been gradually reduced because of its toxicity, it has found its new use in cadmium telluride solar panels because of its capability of absorbing more sunlight than the existing silicon.

When comparing the solar power costs with the costs of fossil fuel and nuclear power, we mainly address the *fixed cost*[9] of power generation. The latest data on future power generation technologies show that solar PV power generation is already among the least expensive options. Solar power generation only has fixed variable operation and maintenance costs but does not have variable operation and maintenance costs or *marginal costs*,[10] that is, the cost of fuel.

This means that solar PV power generation, like wind power generation, only needs to acquire and maintain equipment but not fuel because both solar and wind energy used for power generation are freely available. In addition, another cost associated with fossil fuel and nuclear power generation needs to be considered in the comparison. This is the so-called *external cost*[11] of dirty power generation environmental and human health effects: emissions of CO_2, sulfur dioxide (SO_2), nitrogen oxides (NO_X), mercury (Hg), and radiating waste.

In addition, smaller solar PV power plants or systems normally have higher capital costs, approximately 10 ct/kWh for rooftop installations of a few kW nominal capacities. Older PV power plants produce solar electricity much more expensively due to the previously very high investment costs.

Further cost reduction. Although the cost of solar PV power has significantly decreased, further cost reduction is necessary to allow solar PV technology to become affordable to everyone and be fully penetrated everywhere on Earth.

Currently, solar PV technology costs vary significantly across countries. In some Asian countries, such as China, the government FIT policy helped utility-scale solar power installations expand and drastically reduce their PV power costs. As a result, the cost of distributed solar PV systems has already been at or below the coal-fired grid power consumer price and has reached so-called *solar grid parity* in China. PV power generation is already the cheapest technological choice in China and Vietnam. Based on solar grid parity, China's new strategy of encouraging distributed solar is expected to further lower solar PV installation costs and increase solar power penetration in the country[12] (Fig. 1.13).

[9] *Fixed cost*, also called capital cost, is the cost of investment in the production equipment, here the power generation system.

[10] *Marginal cost* is the change in total production cost that comes from making or producing one additional unit, which is generating one additional watt of power.

[11] *External cost* is the theoretical cost that fossil fuel and nuclear power generation and consumption incur to the environment and society, but the power generator and consumer do not pay. Such costs include, for example, environmental cleaning costs and medical costs of respiratory diseases and deaths.

[12] For more information on comparing the future costs of solar PV systems and with those of other power generation technologies, see **Appendix C**.

Power Generation Technology	Capital cost (2020 $/kW)	Variable operating & maintenance (2020 $/MWh)	Fixed operating & maintenance (2020 $/kW-yr)
Combustion turbine—industrial frame	709	4.52	7.04
Combined-cycle—multishift	957	1.88	12.26
Combined-cycle—single shaft	1082	2.56	14.17
Battery storage	1165	0	24.93
Combustion turbine	1169	4.72	16.38
Solar photovoltaic (PV) with tracking	1248	0	15.33
Distributed generation—base	1560	8.65	19.46
Municipal solid waste—landfill gas	1566	6.23	20.2
Solar PV with storage	1612	0	32.33
Internal combustion engine	1813	5.72	35.34
Wind	1846	0	26.47

Fig. 1.13 New central station electricity generating technology costs. *Data Source* NREL

In countries that have not reached solar grid parity, further solar PV penetration needs continued government financial and strategic support to help further lower solar power generation costs. While the U.S. Congress passed legislation providing incentives for distributed solar installations, the U.S. Department of Energy continued conducting research on how to reduce the higher soft costs for residential solar power installations than for utility-scale and commercial solar power installations. The Solar Energy Industries Association called on the federal government for support in reducing the red tape and barriers that slowed the soft cost decline in residential solar power. Soft costs that need to be reduced include federal solar import tariffs, state and local fees for permitting, inspection, and taxes. Additional actions from the federal, state and local governments include establishing online permission and standardization across jurisdictions, reducing wait times and fees, providing financial incentives, and encouraging group purchasing campaigns. In addition to government financial incentives, commercial banks can also help residents and businesses reduce the financial burdens of investment through loan contracts that allow them to pay off the investment over time in installments.

Japan's 2030 solar power target will help reduce solar costs. It supports the deployment of rooftop solar power initiatives: adding 6 GWs by installing solar panels on 50% of central government and municipality buildings; adding 4 GWs by utilizing promotion areas and public land in more than 1000 cities and towns; and adding 10 GWs by increasing the use of solar on parking garages and corporate buildings.

2. Expanding Distributed Rooftop Solar

With the rapid drop in solar PV costs in recent years, solar PV power can be cheaply generated by utility-scale solar power plants in many countries, which will be even more so with further economies of scale in the future. However, solar PV power can also be produced anywhere in a decentralized way, such as by residents and businesses on their own or rented rooftops.

Although small-scale decentralized solar power normally costs more than utility-scale solar power, it also has many advantages. First, it can avoid power losses that would occur with utility-scale solar power where the power has to use distant transmission and distribution grids to reach final power users. This advantage is achieved by the direct use of onsite solar power from residents' and businesses' own or rented rooftops or properties.

Second, as solar panels have become significantly cheaper, making them more accessible for residents and businesses will greatly help further reduce solar power costs and solar power penetration around the world. A new study found that solarizing 50% of the world's rooftops will meet the global annual power consumption. Many countries have realized this great potential and have started to capitalize on it through distributed solar PV power instead of merely relying on utility-scale solar power generation.

Therefore, expanding distributed solar power through rooftop PV installations is considered one of the main solutions to many challenges of solar PV power penetration, such as meeting the higher energy demand and the accelerated net-zero requirement to combat global warming and climate change.

In Germany, solar energy transformation started with distributed rooftop solar power. The German Renewable Energy Act granted higher FIT rates for solar power generated by rooftop solar systems than for solar power generated by utility-scale solar systems. As a result, German solar PV power has been mostly distributed, rooftop-mounted, and owned by households, businesses, and farmers.

The shortage of land for large-scale renewable energy deployment and the lack of urban penetration of renewable energy urged the German government to deliver innovative approaches to generating solar power in cities. The tenant power initiative was one such program designed to encourage landlords to install solar panels and offer their tenants cheap, locally produced power. A Fraunhofer study of the solar potential in Berlin found that using the city's available rooftop space could power up to 25% of city-wide power needs. Accordingly, the city created its master plan to achieve this goal by 2050. Cities can better harness their respective solar potential by creating solar maps and introducing supportive policies that create the right enabling environments for investment.

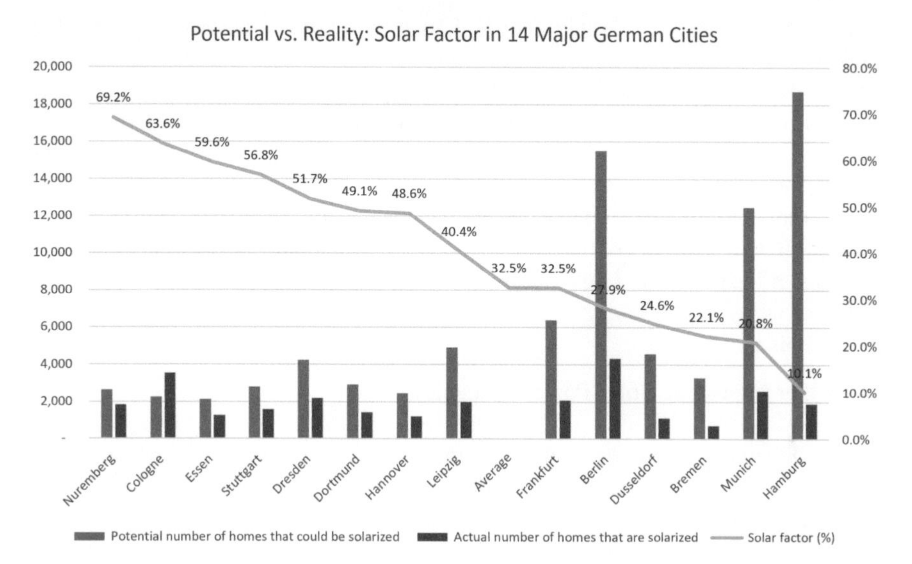

Fig. 1.14 Solar factor: rate of rooftop PV on German new buildings. *Data Source* SolarCheck 2022 by *Lichtblick*

The degree of German cities' move to solar PV power was tracked by LichtBlick, a green power provider in Germany, using the "solar factor" of the cities[13] (Fig. 1.14). Solar factor data indicate that among Germany's 14 most populated cities, five (Nuremberg, Cologne, Essen, Stuttgart, Dresden) managed to harness solar PV power on more than half of their new buildings' available rooftop potential (69.2%, 63.6%, 59.6%, 56.8%, and 51.7%, respectively) by the end of 2020. However, the largest German cities, Berlin and Hamburg, are among the cities that have the lowest new building solar factors. To address this issue, both cities mandated installing rooftop solar PV equipment on all new buildings by 2023. Government mandates of rooftop PV installations on new buildings in other countries, such as those in German cities, would further strengthen the distributed solar power and cost reduction of rooftop solar PV systems in cities around the world.

For countries with larger land areas, distributed solar power penetration is also the key for solar power penetration. The U.S. Department of Energy's Solar Futures Study released in October 2021 and the Solar Energy Industries Association's "30 × 30 analysis" both concluded that over the next decade, distributed solar power must grow between two and four times faster than in the previous decade to reach the nation's net-zero and renewable energy goals at the lowest cost.

[13] *Solar factor* is the ratio of a city's roof-top areas used for solar power generation and its new buildings' total roof-top areas.

In China, which has already achieved solar grid parity, the government's new strategy focuses on encouraging distributed on-site rooftop solar PV installations. This policy deviates from its massive FIT policy that favored utility-scale solar power investment in remote solar-rich deserts. The new solar power strategy is expected to help onsite prosumer solar power generation and self-consumption, reduce long-distance power transmission, relieve the peak demand on the grid during the daytime, and improve the solar power generation efficiency by energy storage. It will also help solar PV technology accelerate its penetration on the rooftops of business buildings, public buildings, and rural homes.

B. Technical Solutions

1. Further Improving Solar Power Efficiency

Despite the notable improvement in the efficiency of solar panels, further improvement of solar cell efficiency is needed to make solar PV power generation more sustainable and environmentally friendly. Improved solar cell efficiency will not only effectively improve citizens' and governments' perceptions of solar power efficiency and their willingness and affordability to pay for solar PV power but also accelerate solar power penetration.

Solar cell efficiency improvement is one of the most significant tasks for scientists and engineers in solar cell R&D. Increased efficiency can reduce the demand for solar panels to provide the same level of power and reduce solar costs for solar panel manufacturing, marketing, installation, and investment. Efficiency improvements will allow solar PV manufacturing to use fewer resources, solar installation and generation to take up less space, and the solar PV industry to reduce its environmental impacts, such as habitat loss for both plants and wildlife.

R&D experiments introducing diamond wire, for example, help make solar cells thinner and less expensive to produce and transport. Passivated emitter rear cell technology is another innovation that converts more solar light into power by reflecting unabsorbed light.

There are two new types of solar cells in research labs—multijunction silicon solar cells and thin film solar cells. Most of these new solar cells have reached efficiencies similar to those of the monocrystalline or polycrystalline solar panels on the market. Some of them, such as perovskite cells, have even achieved higher efficiencies than marketed cells.

Perovskite is a mineral composed of calcium, titanium, and oxygen. The solar efficiency of perovskite cells is significantly improved because perovskite panels can be manufactured as very thin layers, require much less material, and use a much less energy-intensive process than silicon cells. By placing perovskites on top of the silicon layer, the resulting tandem cell's[14] efficiency rate reaches 30%, which is better than those of the best available silicon panels. Much higher

[14] Tandem cells refer to stacks of p-n junctions, each of which is formed from a semiconductor of different bandgap energy.

efficiency is achieved by the perovskite tandem cell's ability to capture different bandgaps of sunlight, such as absorbing infrared light, which current silicon solar cells are unable to capture.

However, the new solar cells still have significant weaknesses and need further improvements to meet market needs. To be commercially viable, multijunction solar cells need to significantly reduce their manufacturing costs, and thin film solar cells need to resolve the issues of conversion rate, toxicity, instability, and short durability/longevity; that is, they need to become safe and stable enough to survive 20 years outdoors. At the same time, both multijunction solar cells and thin film solar panels need large-scale, low-cost manufacturing techniques to become affordable for the solar power market.[15]

Thin-film solar panels still have lower conversion rates than crystalline silicon panels, and ongoing research and development efforts have focused on enhancing their efficiency. Advances in materials, device structures, and manufacturing processes are gradually improving the conversion rates of thin-film technologies. As these efficiency improvements continue, the cost-effectiveness of thin-film solar panels will further strengthen.

2. Increased Flexibility

Despite the lower conversion rates of thin-film solar cells, thin-film solar panels can be seen as potentially more cost-effective options because of their reduced production cost, increased flexibility, and lower weight for installation.

Thin-film solar panels require significantly less material than crystalline silicon panels. They are made by depositing a thin layer of photovoltaic material, such as amorphous silicon (a-Si), cadmium telluride (CdTe), or copper indium gallium selenide (CIGS), onto a substrate. This reduced material usage can lead to lower production costs and potentially lower material costs in the future as thin-film technologies continue to advance.

Less demanding high-throughput, continuous manufacturing processes such as roll-to-roll or sheet-to-sheet manufacturing also offer advantages in terms of scalability and automation. In addition, the deposition techniques used in thin-film manufacturing can be less energy intensive than the crystal growth processes involved in traditional silicon wafer production. Thin-film solar panels are typically more flexible and lightweight than crystalline silicon panels. These features allow thin-film panels to be more effectively integrated into a wider range of applications, including curved or vertical surfaces such as façades in building-integrated photovoltaics (BIPV). While the advantages of scalable and automated manufacturing of thin-film cells can achieve cost reduction in production, the lightweight design also reduces costs in installation and structural support.

[15] For more information on research solar cells' efficiencies, see **Appendix A**.

New types of thin-film solar technologies, such as perovskite solar cells, show promising potential for achieving high conversion efficiencies at lower manufacturing costs. Perovskite solar cells are relatively simple to manufacture and can be deposited on various substrates, including flexible materials. Although these technologies are still in the research and development stage, they hold great promise for future cost-effective solar PV deployment.

As technology continues to advance and manufacturing processes become more optimized, thin-film solar panels have the potential to play a significant role in the future deployment of solar PV systems, providing a cost-effective alternative to traditional crystalline silicon panels.

3. Floating Solar

In addition to being ground-mounted and rooftop-mounted, solar PV systems can also be mounted on a structure that floats on a body of water, such as a hydropower reservoir, lake, or sea. On-water solar power has several major advantages over ground-top or rooftop solar power. First, as a utility-scale solar power technology, floating solar can avoid land use challenges facing ground-mounted solar power plants. Second, on-water silicon PV panels increase power output by 5–10% over their ground-top or rooftop counterparts because of the cooling effect of water on them. Third, it complements existing hydropower generation; covering only 3–4% of the reservoir with floating solar panels will double the dam's existing power generation capacity. In addition, floating solar water coverage has several additional environmental benefits, such as reducing water evaporation, controlling blue algae, and cooling water.

4. Adding and Revolutionizing Energy Storage

Because of the high costs of both solar power and power storage, as well as the generous government FIT support for grid-connected solar, the inclusion of investment in power storage was initially not considered in most solar power projects around the world. However, when intermittent and variable solar power has reached higher levels of penetration and distributed rooftop solar power has taken a rising share and relevancy in the total power generation mix, increasing governments have reduced or phased-out FIT incentives for solar power. Adding power storage to solar PV systems has become increasingly important.

At higher penetration levels of distributed solar PV power, power storage can help with the flexibility and efficiency of solar power generation through after-dark self-consumption. Energy storage is becoming a key part of globally distributed rooftop solar PV markets. The widespread use of energy storage for rooftop installations will help substantially reduce energy storage costs. Adding substantially more energy storage capacity to utility-scale and distributed solar systems and the grid will help strengthen the grid stability, power quality, efficiency, and affordability of solar PV power.

Power storage is paramount in areas with the most productive solar power generation and curtailment. Using surplus solar power to generate green hydrogen can reduce the waste of power caused by curtailment and contribute to cleaning up the currently dominant "gray" hydrogen (96% of global hydrogen production) made of fossil fuels—coal, oil, and natural gas. Transforming it to a major form of energy storage for green transportation and grid regulation in addition to pumped hydropower storage will also help increase flexibility and safety to the future 100% renewable energy grid. The widening price gap between cheap solar and more expensive coal in solar-rich areas will allow solar system prosumers and businesses to afford additional lithium-ion battery storage at a price below that of coal-fired power to increase the efficiency of their solar systems.

Breakthroughs in energy storage technology hold great promise in solar power penetration and the fight against climate change. Improving current power storage technology and advancing next-generation technology will allow the integration of solar and other renewable power sources and achieve Net Zero goals in the future. The U.S. Better Energy Storage Technology Act is expected to provide U.S. research initiatives to achieve innovative breakthroughs in renewable energy storage technologies aimed at reducing energy storage costs by 90% in the next decade.

5. Improving and Building Renewable Energy Grids

Further solar power penetration requires effective transmission and distribution grids. Long-distance transmission of solar power has been upgraded to *ultrahigh-voltage direct-current (UHVDC) or ultrahigh-voltage alternating-current (UHVAC)*[16] power grids in China, India, and Brazil, but further improvements are needed through better coordination of decentralized power grid management and better communication between the transmission grids and the distributing grids to further reduce the highest solar curtailment rates in the Northeast and Northwest regions and better utilize the best solar resources in those regions.

In the U.S., grid improvements and innovations are also paramount for the transmission of solar power across large distances, for example, from southern California and Texas, where it is sunny, to the cloudy northeast. To accommodate the ever-growing share of solar power and other renewable power sources, smart grid technology needs to be developed and deployed. To improve solar PV power grid integration, the U.S. Department of Energy announced funding of $45 million for research to advance solar hardware and systems integration, including

[16] Ultra high voltage (UHV) transmission is associated with $\geq \pm 800$ kV ultrahigh-voltage direct current (UHVDC) grids and ≥ 1000 kV ultrahigh-voltage alternating current (UHVAC) grids, which are needed for transmitting large amount of power over a long distance without significant power loss.

the creation of a consortium dedicated to developing control technologies for a modernized electric grid. Smart grid technology, including artificial intelligence (AI) grid monitoring, communication, and management, is essential to make grid operation more effective.

6. Space-Based Solar Power

Space-based solar power involves using satellites equipped with solar panels to capture solar energy in space and transmit it wirelessly to Earth. This method offers advantages such as uninterrupted energy collection and the ability to capture more solar radiation compared to terrestrial panels. The energy can be transmitted as microwave or laser beams. Microwave-transmitting satellites would operate in geostationary orbit approximately 35,000 km away from Earth, while laser-transmitting satellites would be closer at approximately 400 km. Laser satellites have lower startup costs and simpler ground implementation but face safety concerns and difficulties in transmitting power through clouds. Each satellite would have a relatively low power output of 1–10 MW, requiring multiple satellites to have a significant impact. Challenges include high production costs, space-based assembly, and the difficulty of repairs due to the satellite's distance from Earth.

Summary

Solar PV power is a renewable energy technology that is used to generate power by converting sunlight into electricity. This technology differs from power generation technologies using fossil fuels because the energy source sunlight it uses is the cleanest and safest; solar PV power generation has the lowest CO_2 emissions and the lowest disease and death rates among all power generation technologies. It has great potential to offset CO_2 emissions from fossil fuel-fired power generation and meet the global Net Zero target. However, solar PV power generation still has many challenges and urgently needs solutions for its full penetration.

There is no single panacea that could solve the multiple challenges to solar PV power. To reduce the high upfront solar power investment, several options need to be taken into consideration. For example, educating consumers on price comparison with fossil fuel-fired power in terms of avoided fuel cost and external cost; providing government funding for new solar cell R&D; expanding distributed rooftop solar power generation, reducing red tape and related soft costs, providing solar power investors government financial incentives or commercial loan programs can help residents and businesses pay off investment over a longer period.

In the current solar PV market, most solar modules are made of two types of silicon cells—polycrystalline silicon cells and monocrystalline cells. The former is more efficient in converting sunlight into power but is also more expensive than the latter. The efficiency of silicon solar panels on the market ranges between 20 and 26%. New solar cells that are intensively researched and developed in labs include multijunction solar cells and thin film solar cells. Multijunction solar cells have reached 46% high efficiency, but their complicated and costly manufacturing makes them unviable for the solar market. Currently, they are only used for satellites and military drones because of their high costs. The efficiency of thin film solar cells has become similar to that of silicon cells, and the efficiency of some types of thin film cells, such as perovskite solar cells, even exceeds that of the best silicon cells. However, before they can be mass-produced and marketed, multijunction cells need to reach a much lower price, and perovskite cells need much higher stability, durability/longevity, much safer approaches to toxicity, and mass manufacturing techniques.

PV power generation requires sunlight. Any natural or physical conditions, such as nights, clouds, smog, shorter days, or shade, which prevent sunlight from hitting solar modules will prevent or reduce power generation and cause instability and fluctuation in the grid. In addition, the uneven geographic distribution of solar resources also causes a mismatch between solar power generation and consumption, which in turn leads to power oversupply and curtailment in solar-rich but economically underdeveloped areas and power outages in economically developed power-hungry cities.

Options for solving these solar power supply and demand challenges include building long-distance power transmission grids using high-voltage, ultrahigh-voltage DC, and AC grid technologies to connect power output and power load centers and using smart grid technology to make grid operation more effective. It is also vital to encourage rooftop and water-top solar power installation and consumption to upscale solar power installation, generation, and consumption.

The market conditions need to be improved for utility-scale solar power generation for selling power and distributed rooftop solar power generation for self-consumption. The former's efficiency can be improved by adding grid-scale energy storage. The latter's advantages, such as avoiding transmission and distribution costs, can also be further strengthened by adding less energy storage. With added energy storage, the efficiency of the existing solar power equipment can be improved by storing mid-day solar glut and self-consuming the stored solar power to meet the evening peak demand.

Activities

Further Readings

1. United States Environmental Protection Agency. (2022). Greenhouse Gas Emissions from the Electricity Sector. https://www.epa.gov/ghgemissions/sources-greenhouse-gas-emissions#Electricity-Sector
2. Bailie, C. (2015). How to make solar power more efficient and affordable? World Economic Forum. https://www.weforum.org/agenda/2015/03/how-to-make-solar-power-more-efficient-and-affordable

3. Wang, L. & Wang, M. (2022). China's Curtailments go up as Renewables Growth Explodes. https://www.solarpaces.org/chinas-curtailments-go-up-as-renewables-growth-explodes/
4. Kazmeyer, M. (2018). Future of Solar Power: Obstacles & Problems. https://sciencing.com/future-solar-power-obstacles-problems-21852.html
5. Solar Energy Industry Association (2023). Solar Industry Research Data: Solar Industry Growing at a Record Pace. https://www.seia.org/solar-industry-research-data
6. International Energy Agency (2019). Distributed Solar PV. https://www.iea.org/reports/renewables-2019/distributed-solar-pv
7. International Energy Agency. (2021). Energy Technology Perspectives 2020. https://www.iea.org/reports/energy-technology-perspectives-2020
8. Sovacool, B. K., & Dworkin, M. H. (2014). Global electricity mortality risks and external costs: new evidence from 2012 air pollution and meteorological data. Environmental Research Letters, 9(8), 084,010. https://iopscience.iop.org/article/10.1088/1748-9326/9/8/084010
9. Wang, R., Hasanefendic, S., Von Hauff, E., Bossink, B. (2022). The cost of photovoltaics: Re-evaluating grid parity for PV systems in China. Renewable Energy, 194, 469–481. https://doi.org/10.1016/j.renene.2022.05.101
10. Sendy, A. (2021). What are thin film solar panels, how do they work and why aren't they used for residential solar systems? https://www.solarreviews.com/blog/thin-film-solar-panels
11. Hockenos, P. (2022). Will Solar Mandates Prompt a Boom in Europe's Rooftop Solar? https://energytransition.org/2022/12/will-solar-mandates-prompt-a-boom-in-europes-rooftop-solar/
12. Yang, P. (2022). Urban expansion of *Energiewende* in Germany. https://doi.org/10.1186/s13705-022-00373-1
13. Akshay, V.R. (2023). The Future of Solar Energy: Predictions for 2023 and Beyond. https://thesolarlabs.com/ros/future-of-solar-energy-predictions-for-2023/
14. O'Malley, I. (2023). The first generation of solar panels will wear out. A recycling industry is taking shape. https://www.msn.com/en-us/money/companies/the-first-generation-of-solar-panels-will-wear-out-a-recycling-industry-is-taking-shape/ar-AA1eDvsr

Closed Questions

1. What is the best answer for "What is solar power?"
   (a) Electric energy generated from solar heat
   (b) Solar photovoltaic power
   (c) Concentrated solar power
   (d) Utility-scale or distributed solar PV
   (e) All the above.

2. What is not solar power?
   (a) Electric energy generated from solar light
   (b) Electric energy generated from solar heat
   (c) Thermal energy generated from solar heat
   (d) Solar photovoltaic power
   (e) Concentrated solar power.
3. What is solar PV power, and how does it work?
   (a) A technology that harnesses wind to generate electricity
   (b) A method of converting solar energy into heat for heating purposes
   (c) A technology that converts sunlight directly into power using photovoltaic cells
   (d) Process of using water currents to generate electrical power
   (e) A method of utilizing geothermal heat for electricity production.
4. What are the key components of a solar PV system?
   (a) Turbines, generators, and transmission lines
   (b) Solar panels, inverters, and batteries
   (c) Coal-fired boilers, steam turbines, and cooling towers
   (d) Wind turbines, transformers, and grid connections
   (e) Solar concentrators, receivers, and heat exchangers.
5. How is solar PV power different from other renewable energy sources?
   (a) Solar PV power does not rely on any natural resources
   (b) Solar PV power is the only source that produces heat and electricity simultaneously
   (c) Solar PV power is the most expensive renewable energy option available
   (d) Solar PV power is the only source that produces power continuously
   (e) Solar PV power directly converts sunlight into electricity using photovoltaic cells.
6. What are the main benefits of solar PV power?
   (a) It creates job opportunities in the fossil fuel industry
   (b) It produces harmful emissions that contribute to air pollution
   (c) It provides a consistent and reliable source of electricity
   (d) It reduces greenhouse gas emissions and helps combat climate change
   (e) It requires large amounts of water for its operation.
7. How has the cost of solar PV technology evolved over the years?
   (a) It has rapidly decreased, making it more affordable
   (b) It has remained constant without any significant changes
   (c) It has steadily increased due to high demand
   (d) It has experienced occasional fluctuations but remains high
   (e) It has only increased due to the limited availability of materials.
8. What is a solar cell?
   (a) A basic component in a solar thermal collector
   (b) A type of electric energy storage, such as a battery
   (c) A basic component in a solar tower that converts solar heat to power

(d) A basic component in a solar panel or installation that converts solar light to power
(e) None of the above.

9. What is the most significant benefit of solar energy?
(a) It is one of the densest energy sources
(b) It is one of the cleanest and safest energy sources
(c) It is available anytime at a given location
(d) It is the most mature power generation technology that does not need further innovation
(e) All the above.

10. What is the largest hurdle to solar PV power generation?
(a) High operation and maintenance costs
(b) High levelized cost of energy
(c) High upfront capital cost
(d) High fuel cost
(e) All the above.

11. What is not a challenge of solar PV power generation?
(a) Low solar heat
(b) Lack of land or surface areas for solar PV installations
(c) Persistently cloudy or rainy days
(d) Lack of supply chain and service
(e) Lack of funds for investment.

12. What causes solar power curtailment?
(a) High local consumption of solar PV power
(b) High rate of energy storage of solar PV power
(c) Excessive solar PV power generation on sunny days
(d) Low solar PV power generation on cloudy days
(e) Absence of local congestion of solar PV power generation.

13. What is not a solution for the high upfront capital cost issue for solar PV power?
(a) Mandating solar PV installations in new buildings
(b) Requiring the 100% local manufacturing of solar PV installations
(c) Further improving the efficiency of solar cells
(d) Providing government financial support for R&D of solar power generation
(e) Upscaling distributed PV power installations in residential and business areas.

14. What are the market dynamics driving the growth of solar PV power globally?
(a) Decreasing government support and incentives
(b) High cost compared to other energy sources
(c) Increasing public awareness and demand for renewable energy
(d) Limited availability of solar PV technology
(e) Volatile and unpredictable energy prices.

15. What are the key challenges associated with the widespread adoption of solar PV power?
    (a) Limited sunlight availability in most regions
    (b) Difficulty in integrating solar PV with existing power grids
    (c) High maintenance costs of solar PV systems
    (d) Lack of technological advancements in solar PV technology
    (e) Insufficient manufacturing capacity for solar PV components.
16. How does solar PV power contribute to reducing carbon emissions?
    (a) By emitting harmful greenhouse gases during operation
    (b) By relying on fossil fuels for electricity generation
    (c) By reducing the need for traditional fossil fuel-based power plants
    (d) By increasing carbon emissions due to manufacturing processes
    (e) By consuming excessive energy during the manufacturing stage.
17. What is the potential of solar PV power in meeting global energy demand?
    (a) Solar PV power can only provide a small fraction of global energy needs
    (b) Solar PV power has the potential to meet a significant portion of global energy demand
    (c) Solar PV power is not a reliable source for meeting energy demand
    (d) Solar PV power can only be used in specific geographical locations
    (e) Solar PV power has no potential in meeting global energy demand.
18. What innovative solar PV technologies are being developed?
    (a) Traditional silicon-based solar panels
    (b) Concentrated solar power (CSP) systems
    (c) Solar PV systems with no storage capabilities
    (d) Thin-film solar cells using organic materials
    (e) Solar PV technologies that are not currently under development.
19. What role does government policy play in promoting solar PV power?
    (a) Government policies have no impact on the promotion of solar PV power
    (b) Government policies discourage the use of renewable energy sources
    (c) Government policies focus solely on supporting nonrenewable energy industries
    (d) Government policies increase taxes on solar PV equipment and installations
    (e) Government policies provide subsidies and incentives for solar PV installations.
20. How does solar PV power integrate into the existing electricity grid?
    (a) Solar PV power cannot be integrated into the existing grid system
    (b) Solar PV power is easily integrated with the existing grid through inverters and interconnections
    (c) Solar PV power requires the complete overhaul of the electricity grid infrastructure
    (d) Solar PV power only works in isolated, off-grid systems
    (e) Solar PV power requires a separate grid for its operation.

Open Questions

1. What is solar PV power? How does it differ from power generation using fossil fuels?
2. Why is solar power paramount for combating climate change and global warming?
3. What types of solar PV cells are there? What are the advantages and disadvantages of the different solar PV cell types?
4. What is the range of power generation efficiency or the conversion rates of the current solar PV cells?
5. What are the requirements for solar PV power? What challenges do these requirements pose?
6. What are the efficiencies or conversion rates of the current and new solar cells?
7. What is the potential of the new solar cells for the future solar power market?
8. What are the hardware costs and soft costs of solar PV power installation?
9. Why do soft costs have a higher share than hardware costs?
10. What options are there to reduce soft costs?
11. What challenges do solar PV power generation?
12. What solutions are available to address these challenges?
13. What is solar power curtailment?
14. What are the causes of this curtailment problem?
15. What are the possible solutions to this problem?
16. What is utility-scale solar power?
17. What are the benefits and disadvantages of this form of power generation?
18. What are the solutions to the challenges of this form of power generation?
19. What is distributed solar power?
20. What are the benefits and disadvantages of distributed solar power generation?
21. What are the solutions to the challenges of this form of power generation?
22. How does solar PV power integrate into the existing electricity grid?
23. What are the environmental impacts of solar PV power?
24. Can solar PV power be used in off-grid or remote areas?
25. What are the main factors influencing the efficiency of solar PV panels?
26. How does solar PV power compare to other sources of electricity in terms of reliability?
27. What are the different types of solar PV installations, such as rooftop and utility-scale installations?
28. Are there any specific challenges related to the storage of solar PV power?
29. How do solar PV power systems handle variations in sunlight intensity and weather conditions?
30. What are some future trends and developments expected in the solar PV power industry?

2 Solar Thermal Energy

Introduction

Science and Technology

1. Solar Thermal Energy

Solar thermal energy is a type of renewable energy harnessed from sunlight by solar thermal technologies. Solar thermal technology can be divided into two groups: concentrated solar power generation and solar heat applications.

For solar heat applications and concentrated power generation, solar heat is classified as low-temperature heat, medium-temperature heat, or high-temperature heat. Solar heat at different temperatures can be used for different applications. Low-temperature heat can be used for collectors, and heat transfer media are used to collect solar heat at different temperatures for swimming pool heating and low-grade water and space heating. Medium-temperature heat is used for domestic and district hot water and space heating.

Concentrated solar power generation (CSP), industrial processes, solar district heating and cooling (SDHC) system enhancement, and absorption chilling. To harness solar heat at different temperatures, different solar heat technologies must be used. While the collection of solar heat at low and medium temperatures only requires solar heat collectors, the generation of solar heat at elevated temperatures and pressures (e.g., 300 °C/572 °F and 20 bar) requires mirrors (also called lenses or heliostats) in addition to collectors. Figure 2.1 shows the heat temperatures, applications, and technologies of solar heat generation.

2. Concentrated Solar Power

Concentrated solar power (CSP, also referred to as concentrating solar power or solar thermal power) stands for modern technologies that utilize the heat harnessed

P. Yang, *Renewable Energy*, https://doi.org/10.1007/978-3-031-49125-2_2

	Low Temperature	Medium Temperature	High Temperature
Temperature	< 43 °C (110 °F)	> 43 °C (110 °F) Default: 60-82°C (140-180°F)	> 82 °C (180 °F)
Solar heat collector	Unglazed liquid flat-plate collectors Metallic or nonmetallic absorbers	Glazed evacuated tube collectors Glazed flat-plate collectors	Parabolic dish Parabolic trough Lineal Fresnel Solar tower
Application	Swimming pool heating Low-grade water and space heating	Domestic water and space heating District water and space heating	CSP generation Industrial processes SDHC enhancement Absorption chilling
Heat transfer instrument	Water	Water Air	From molten salt to water to steam From oil to water to steam

Fig. 2.1 Solar heat: temperature, collector, application, heat transfer media

from sunlight to generate renewable power. Although both CSP and PV use sunlight to generate power and both belong to solar power, CSP technologies differ significantly from solar PV power generation technology.

While PV technology directly converts the sun's radiation to power by solar PV cells absorbing sunlight, knocking loose electrons, and causing them to flow as direct current (DC) power, CSP technologies generate power using an indirect approach. The first step uses various technologies to reflect, concentrate, and collect sunlight and converts it to high-temperature heat. Once the scorching solar heat is generated, the second step converts it to mechanical energy to drive the steam turbine and electric generator for power generation.

The technologies used in the first step are mirrors or reflectors in various configurations. These configurations of the mirrors or reflectors of CSP give the names of most solar thermal power-generating technologies. There are four main configurations: parabolic trough, parabolic dish, linear Fresnel reflector, and solar tower. Each of these CSP technologies consists of mirrors or lenses (also called concentrators, reflectors, or heliostats) that reflect and concentrate sunlight (photons); a receiver that collects solar heat from the concentrated sunlight; an optional heat storage that stores solar heat and allows longer operating hours of power generation in the absence of sunlight; and a steam turbine and power generator that generate solar power Fig. 2.2.

3. CSP Generating Technologies

Parabolic trough systems are the most common type of CSP technology. These systems use parabolic-shaped (curved) reflectors to concentrate sunlight onto a receiver tube running along the focal line of the trough. A heat-transfer fluid flowing through the receiver absorbs the concentrated solar heat and transfers it to a power block, where it is used to generate steam and drive a turbine. The round-trip efficiencies of parabolic trough collectors can range between 60 and 80%. The operating temperatures for these collectors are significantly higher, usually in the

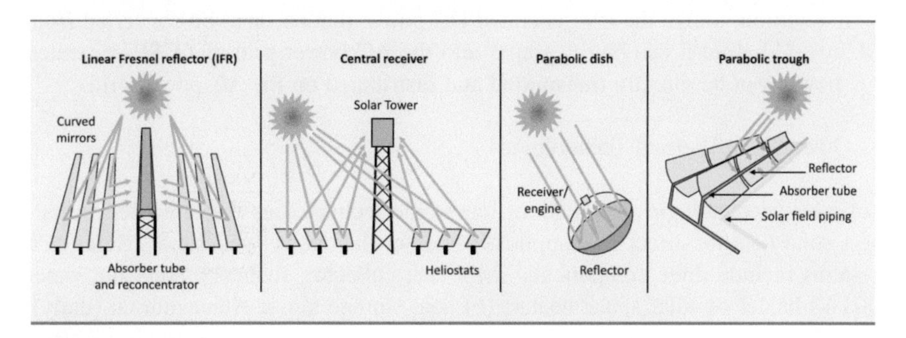

Fig. 2.2 Types of concentrated solar power systems. *Image Credit* NREL

range of 150–400 °C (302–752 °F). The round-trip efficiency of parabolic trough collectors can vary, but it typically falls in the range of 60–70%.

Parabolic dish collectors use a parabolic-shaped dish to concentrate sunlight onto a receiver located at the focal point. These collectors offer high efficiency and are often used for applications such as solar cooking and small-scale power generation. They can achieve heat at very high temperatures, ranging between 500 and 1500 °C (932–2732 °F). The round-trip efficiency of parabolic dish collectors can reach 80% or more.

Dish-Stirling systems consist of a parabolic dish that concentrates sunlight onto a receiver at the focal point. The receiver contains a Stirling engine, which converts the heat energy into mechanical power and then drives a generator to produce power. These systems can achieve output temperatures between 250 and 1000 °C (482–1832 °F). Because of the technological complexity and related high costs, the round-trip efficiency of dish stirring systems can range from 40 to 50%, lower than that of parabolic dish systems.

Solar power tower systems, also referred to as central receiver or power tower systems, use a field of mirrors called heliostats to concentrate sunlight onto a central receiver mounted on top of a tower. The receiver absorbs the concentrated solar heat and transfers it to a working fluid, such as molten salt, which is used to generate steam and drive a turbine. These systems can reach extremely high output temperatures beyond 1000 °C (1832 °F), often reaching up to 1500 °C (2732 °F) or more.

Different types of concentrators produce different peak temperatures and correspondingly varying thermodynamic efficiencies based on the differences in the way they track the sun and focus sunlight. The concentrated solar heat is harnessed through a heat receiver, which heats a fluid, such as molten salt or oil, in a heat pipe, and the heated fluid can be directly sent to a heat exchanger to boil water to create steam that drives a steam turbine (heat engine) and power generator to generate alternating current (AC) power. However, the surplus heated fluid can also be easily stored if the CSP is equipped with heat storage.

In addition, unlike the PV-generated DC power that needs to be converted from DC to AC before it can be integrated into the AC power grid, the CSP-generated AC power can be directly transmitted and distributed on the AC power grid.

4. Other Solar Thermal Technologies

Solar heat technologies are modern renewable technologies that are used to harness solar heat for direct heat applications other than power generation. Solar heat systems include three components: (a) a heat collector, such as a solar hot water (SHW) heater or solar space heater; (b) heat storage tanks, either diurnal (daily), long duration, or seasonal; and (c) a distribution system that expands within a building structure or a district heating network.

The types of solar collectors include unglazed collectors, glazed flat-plate collectors (GFPCs), evacuated tube collectors (ETCs), concentrated collectors with water as the solar heat carrier, glazed and unglazed air collectors, and photovoltaic thermal (PVT) collectors (Fig. 2.3).

Flat-plate collectors are the most common and widely used type of solar thermal collectors. They consist of a flat, insulated box with a dark absorber plate covered by a transparent glass or plastic cover. The sunlight passes through the transparent cover and is absorbed by the plate, which heats up and transfers the heat to a fluid flowing through tubes or channels within the collector. These collectors are commonly used in solar water heating systems. They typically operate at

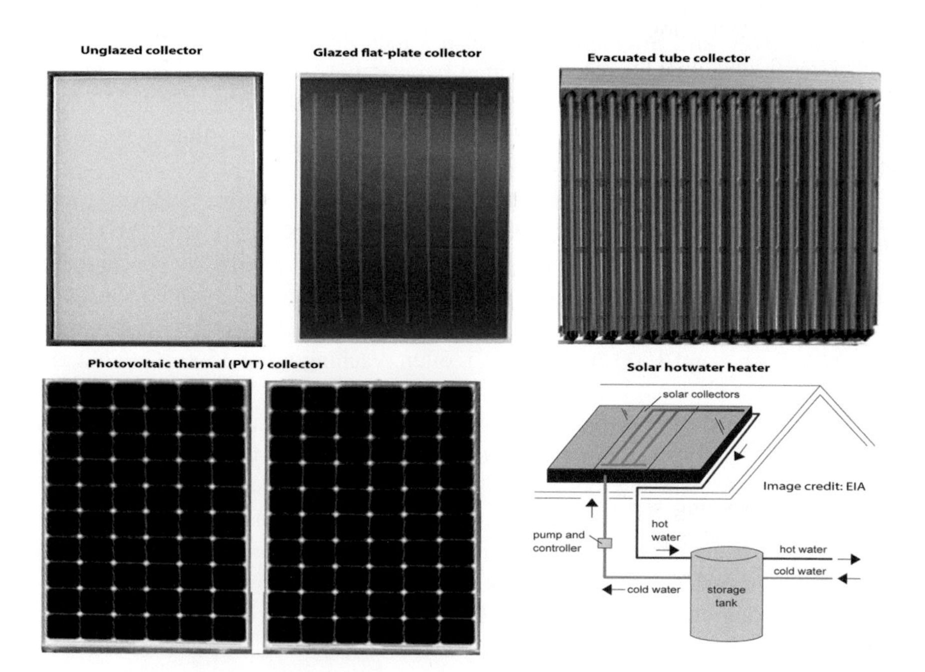

Fig. 2.3 Types of other solar thermal collectors

lower temperatures, ranging from approximately 30–0 °C (86–176 °F). The round-trip efficiency of flat-plate collectors ranges from 40 to 60%, depending on factors such as insulation, glazing, and heat transfer losses.

Evacuated tube collectors are composed of a series of parallel glass tubes, each containing an absorber plate. The air is evacuated from the tubes to create a vacuum, which reduces heat loss through convection and conduction. Sunlight enters the tubes, is absorbed by the plates, and heats up the fluid flowing through them. Evacuated tube collectors are more efficient than flat-plate collectors and are often used in both residential and commercial applications. These collectors achieve higher output temperatures, typically between 60 and 120 °C (140 and 248 °F), and round-trip efficiencies in the range of 50–70%, depending on factors such as the design, insulation, and heat transfer mechanisms.

Solar air collectors consist of an absorber plate, usually made of metal, which absorbs solar radiation and heats up the air flowing through it. Solar air collectors are used to heat air directly but are also commonly used for ventilation and drying applications. The output temperature of solar air collectors typically ranges from approximately 30–80 °C (86–176 °F) or higher, depending on factors such as the intensity of sunlight, collector design, airflow rate, and insulation. The round-trip efficiency of a solar air collector ranges from 30 to 70% depending on factors such as collector design, materials used, insulation, air flow rate, and operating conditions.

Photovoltaic thermal collectors (PVTs) are a modern hybrid type of solar energy technology that converts sunlight into both power and heat by combining PV and solar thermal technologies in a single unit. These systems consist of photovoltaic cells and an integrated heat exchanger. The PV cells convert sunlight directly into power, while the heat exchanger absorbs the excess heat from the PV cells and transfers it to a fluid (usually air or water) for various applications, such as space heating, water heating, or industrial processes. There are two main types of PVT collectors—air-based PVT collectors and liquid-based PVT collectors—based on the heat transfer media they use. Air-based PVT collectors use air as the heat transfer medium, while liquid-based PVT collectors use a mixture of water and glycol as the heat transfer medium. The heat transfer fluids pass through the collectors, absorb the heat from the PV cells, and circulate to deliver the collected heat for space heating, domestic hot water, or other thermal applications.

By combining PV and solar thermal technologies, PVT collectors can achieve higher overall energy conversion efficiencies (ranging from 50 to 80%) compared to standalone PV (approximately 20–25%) or solar thermal systems (approximately 30–70%). As a compact and integrated solution, PVT collectors use less space than operating separate PV and solar thermal installations. The integrated heat exchanger also provides solar cells with a useful cooling effect and helps improve the efficiency and lifespan of solar cells, leading to increased power generation.

While solar hot water supply and solar space heating are the most common thermal applications of the heat harnessed from sunlight, solar heat can also be used for solar cooling (also called solar air cooling) or solar air conditioning (regulating both air temperature and humidity), which is mainly popular in the U.S. and Canada. Solar cooling technology is a heat-driven cooling system called an absorption cooling system (ACS). Similar to the regular compression-based air cooling or air conditioning system (AC) or food refrigerator, the ACS cools down the temperature of the air in the building structure using the evaporation of a liquid refrigerant. The main difference between ACS and AC or refrigeration is that ACS uses low-temperature heat, such as solar heat and industrial waste heat, rather than electric power.

Benefits

While solar PV power generation has gained rapid momentum and is highly efficient for power generation, solar thermal applications, including both CSP and direct solar heat applications, offer a range of advantages for addressing specific energy needs in industrial, agricultural, residential, and commercial sectors. Their ability to provide high-temperature heat, complement solar PV in hybrid systems, and address diverse energy demands justifies their expansion in a holistic renewable energy landscape.

1. Benefits of Solar Heat Applications (SHS)

Solar heating and cooling systems can be integrated into existing district heating and cooling systems. The integration can not only provide more sustainable solutions for space heating and cooling in urban areas than the traditional fossil fuel fired district heating and cooling systems, but also perform more efficient than the indirect solutions the solar PV power systems provide. As a whole, solar heating and cooling can help reduce both energy consumption and carbon emissions in densely populated regions.

Solar heat systems can efficiently provide hot water for residential, commercial, and industrial use. The solar enhanced water heating systems are particularly effective in meeting hot water demands in various settings.

Solar thermal applications can provide the high temperatures necessary for various industrial processes such as metal smelting, cement production, and chemical manufacturing. These industries often require heat at temperatures that CSP might more effectively provide than solar PV power generation.

Solar heat can be used as process heat for agriculture. Agricultural processing applications such as crop drying and greenhouse heating often require consistent and controllable heat, which high-temperature solar heat can more efficiently provide than solar PV power.

Solar heat can also be directly used for desalination and water treatment. Desalination and water treatment processes using solar heat help address water scarcity issues in regions with ample sunlight. These applications can be vital for ensuring access to clean water.

Industries that require steam for various purposes, such as sterilization, can benefit from solar thermal systems designed to efficiently produce high-pressure steam.

In areas with limited access to clean cooking fuels, solar thermal technology can be used to create solar cookers and stoves, providing a sustainable and environmentally friendly cooking solution. Solar thermal-driven absorption cooling systems can also provide air conditioning in regions with high solar radiation, reducing electricity demand during peak cooling periods.

Solar thermal systems can be integrated with waste heat recovery processes in industries, capturing and utilizing heat that would otherwise be wasted.

Solar thermal energy can also contribute to energy storage through methods like molten salt storage. This can provide a unique and valuable approach to balancing energy supply and demand.

Incorporating solar heat applications contributes to diversifying the renewable energy portfolio and facilitating a more comprehensive, long-term energy strategy that meets a range of energy needs beyond power generation.

2. Benefits of Concentrated Solar Power

Despite the lower cost of solar PV power generation, there are multiple solid reasons that justify the expansion of concentrated solar power (CSP). While the initial cost of CSP may be higher compared to solar PV, its unique capabilities in thermal energy storage, grid stability, high-temperature heat generation, and dispatchable power make it a valuable asset in a well-rounded renewable energy portfolio. As technology advances and economies of scale are realized, the expansion of CSP could become increasingly justifiable and competitive in the energy landscape.

One of the key advantages of CSP is its ability to incorporate heat storage systems, allowing for the storage of excess heat. The stored heat can then be used to generate power after sunset or during periods of low sunlight or high demand. This feature provides CSP with dispatchable power capabilities, which can enhance grid stability and reliability, making it a valuable complement to intermittent renewables like solar PV and wind.

CSP systems can achieve high temperatures and, consequently, high thermodynamic efficiency in heat generation and storage, which makes CSP a suitable technology for industrial processes that require high-temperature heat, such as in chemical production or certain manufacturing processes, where solar PV might be less effective.

Although CSP also uses solar energy to generate power, its thermal energy conversion process differs from solar PV's direct power generating process. This difference allows CSP to fulfill this task. A major difference between CSP and PV technologies is their capability of using high-efficiency heat storage or thermal energy storage. This capability makes CSP a partially dispatchable power generator. The addition of heat storage generates added value by shifting solar power generation to periods of peak demand, allowing CSP plants to provide additional power generating capacity and ancillary services and reducing power generating and integration costs.

The dispatchable CSP coupled with heat storage changes the competitive relationship between CSP and PV to be partially complementary and contributes to the higher overall penetration of solar energy in two aspects. First, solar power generation is made possible long after sunset and during cloudy days. Second, the dispatchable CSP provides grid flexibility by its controllable power generation, contributing to greater penetration of solar PV and other variable generation sources such as wind.

As a thermal power technology, CSP has more in common with other power plants using thermal energy sources such as geothermal, nuclear, gas, and coal. CSP plants can be equipped with thermal energy storage, which stores energy in the form of heat (for example, using molten salt). With storage capacity, CSP can more easily store solar heat generated on sunny days. In turn, the stored heat allows the CSP to generate power at night and on cloudy days and supply power according to market needs.

This feature allows CSP to compete with baseload plants and dispatchable generators such as fossil fuel-fired power plants and makes it a dispatchable form of solar power. This technological advantage grants CSP a more favorable position than other variable renewable energy technologies, such as solar PV and wind power, and allows it to serve as an indispensable renewable energy technology to complement variable renewable energy technologies. Dispatchable CSP is especially important in sunny regions where PV already has high penetration because dispatchable CSP can meet the power demand unfilled by PV capacity at night and on cloudy days.

Relying solely on one energy technology, such as solar PV, can create vulnerabilities in the energy system. Introducing a mix of renewable technologies, including CSP, can enhance energy security by mitigating the risks associated with over-reliance on a single source.

While solar PV is clean, the integration of CSP can further reduce the reliance on fossil fuels, as CSP can provide power during periods when solar PV output is low or at night.

CSP's ability to generate power without greenhouse gas emissions contributes to mitigating climate change and reducing air pollution, aligning with global environmental goals. CSP technology can be combined with conventional power generation methods, like natural gas or biomass, to create hybrid power plants. These hybrid systems can provide continuous power output even when solar radiation is limited, enhancing the overall reliability and stability of the energy supply.

Expanding CSP technology contributes to building a diverse and resilient energy infrastructure that can serve future generations. Over the long term, the advantages and benefits of CSP could potentially outweigh initial cost considerations.

Deployment

A. Solar Heat Applications

Over the last two decades, the global solar heat market has achieved impressive advancement. The installed solar heat capacity grew from approximately 56 GW in 2001–522 GW in 2021, with a compounded annual growth rate (CAGR) of 12% (see Fig. 2.4).

The annual addition of global solar heat generating capacity rebounded by 2% in 2021 after the consecutive decline at an annual rate of 11% in the preceding period between 2014 and 2020 (see Fig. 2.5). In contrast to the decline in the global solar heat market, the solar heat market experienced a rapid average annual growth rate of 15.5% in Denmark, China, Germany, and Austria in the same period.

From the beginning of the millennium, China led the world's solar heat market mainly through its large-scale use of solar hot-water collectors, which accounted for approximately 73% of the global solar heat capacity in 2021. The second largest group of solar heat applications, Europe, had a market share of 12%, and the rest of the world represented 16% of the global solar heat market (see Fig. 2.6).

The remarkable contribution of solar heat applications to global renewable energy deployment is evident when the record solar heat capacity of 522 GW in 2021 is compared with those of the other fast-growing renewable power technologies—wind power (837 GW), solar PV (942 GW), geothermal power (16 GW), and CSP (6.6 GW). This comparison shows that solar heat technology is a viable

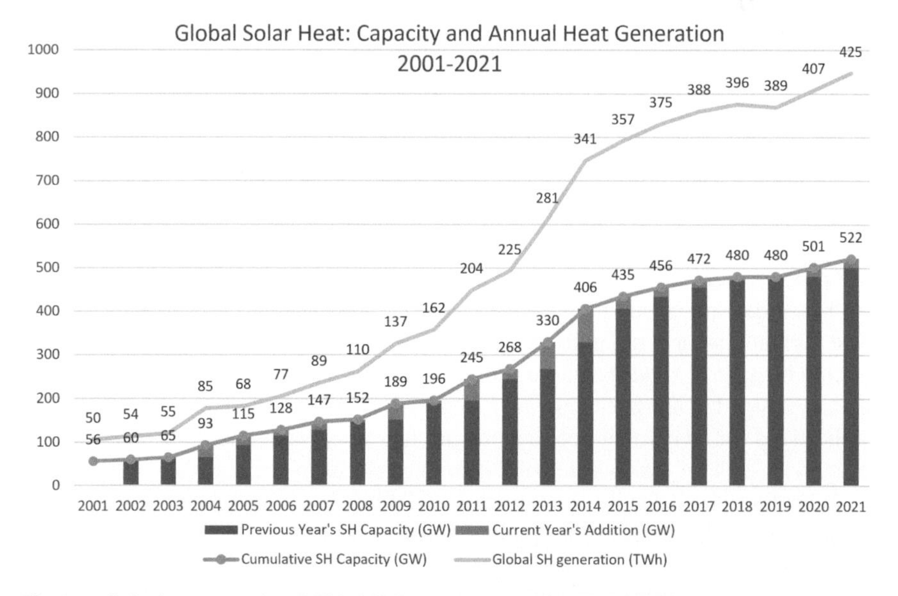

Fig. 2.4 Solar heat expansion (2001–2021). *Data Source* IEA SHC (2023)

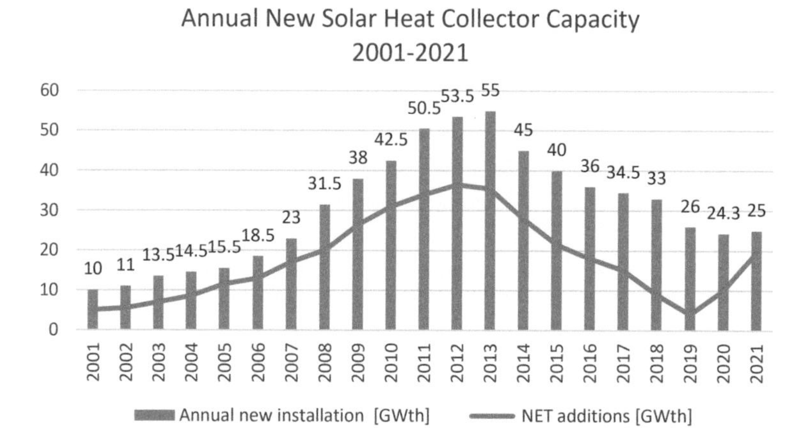

Fig. 2.5 Annual new SH collector capacity. *Data Source* IEA SHC (2023)

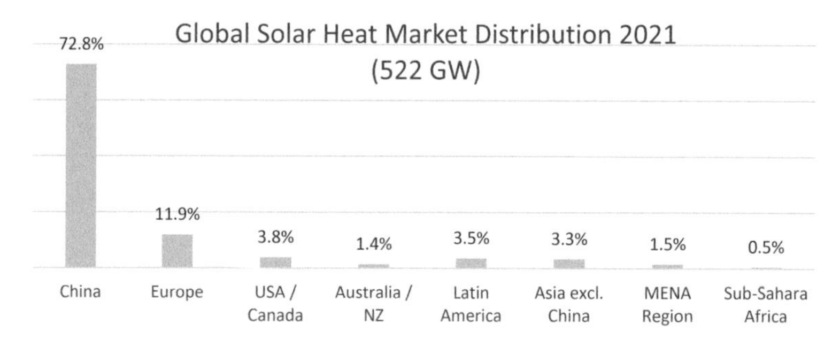

Fig. 2.6 Global solar heat market in 2021. *Data Source* IEA SHC, 2022

and indispensable renewable energy technology that is complementary and competitive with wind power and solar PV power, and it enjoys a significantly stronger market edge over geothermal power and CSP (see Fig. 2.7).

Approximately 88% of global solar heat applications are for hot water supply and space heating in buildings of various sizes. Solar hot-water collectors are the most common solar heat applications in China, whereas solar heat applications for spacing heating in single- and multifamily houses and district heating systems are most developed in Europe, especially Denmark, Germany, and Austria. Solar heat applications for swimming pool heating are more popular in the U.S. and Canada. In general, 88% of large-scale solar heat systems were for hot water supply and space heating in residential, commercial, and public buildings. Solar process heat, which was mainly used in the mining, textile, and food industries, amounted to the remaining 12% (see Fig. 2.8).

Although solar district heating (SDH) systems account for only a small share (2%) of the global solar heat market, these systems play a significant role in the

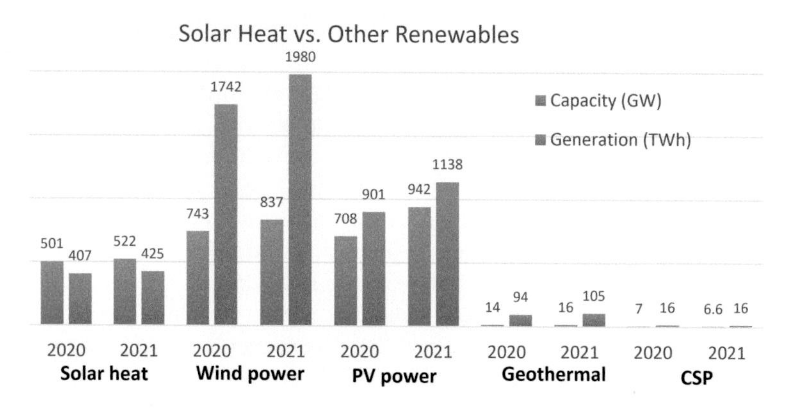

Fig. 2.7 Solar heat compared with renewable power. *Data Source* IEA SHC, 2022

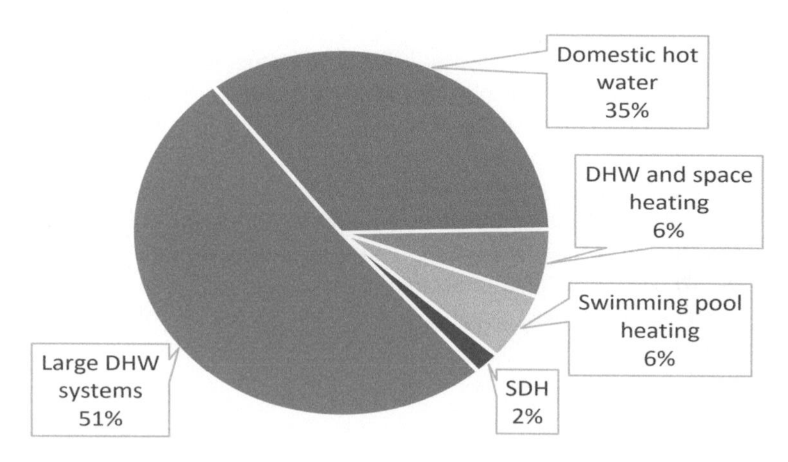

Fig. 2.8 Distribution of global solar heat applications in 2021. *Data Source* IEA SHC, 2022

EU, especially in Denmark, Germany, and Austria. SDH systems provide solar heat to residential, commercial, public, and industrial buildings in the densely populated urban areas of these EU countries. With 130 SDH plants and a combined installed capacity of 1125 MW by the end of 2021, Denmark remained the world's leader in SDH despite its less favorable solar radiation resources compared with those of many other countries, including Germany.

Major players in the SDH market are DESMI, Goteborg Energi, Keppel DHCS, Korea District Heating Corporation, LOGSTOR, Orsted, Shinryo, Statkraft, Arcon-Sunmark, Savosolar, Aalborg, NRG Energy, Alfa Laval, Bosch Thermotechnology Ltd, and Hansen Technologies.

The global SDH market is expected to grow from \$2.52 billion in 2021 to \$2.73 billion in 2022 at a CAGR of 8.14% and to \$3.77 billion in 2026 at a CAGR of 8.42%.

SDH systems are divided into small and large systems. The small district solar heating system refers to installations with small solar collector areas of less than 1000 m^2 and small seasonal thermal energy storage volumes of less than 1000 m^3. SDH systems are widely used in residential, commercial, and industrial applications. Western Europe was the largest region in the SDH market in 2021. The Asia Pacific is expected to be the fastest-growing region in the forecast period. The regions covered in the SDH market report are Asia–Pacific, Western Europe, Eastern Europe, North America, South America, the Middle East, and Africa.

The growing adoption of sustainable energy is driving the growth of the SDH market. Governments, residents, and industries across the globe are focusing on adopting sustainable energy to decrease the dependency on conventional energy consumption and increase the utilization of clean energy such as solar, wind, hydropower, and bioenergy.

Countries actively involved in the SDH market are Australia, Brazil, China, France, Germany, India, Indonesia, Japan, Russia, South Korea, the UK, and the U.S.

The adoption of *photovoltaic thermal* (PVT) collectors is a promising new trend because the market for this type of solar energy collector has gained market interest in recent years. PVT collectors convert solar radiation into both power and heat and thus will play an important role in the energy supply of the future.

Ninety-two percent of solar heat systems in China produce and supply SHW for different users, including multi and single-family buildings. Solar heat systems for space heating in individual buildings and solar district heating (SDH) systems, which refer to large-scale district heating systems equipped with solar collectors as part of the integrated heat supply, have only a small share of 2% in its solar heat market (see Fig. 2.9).

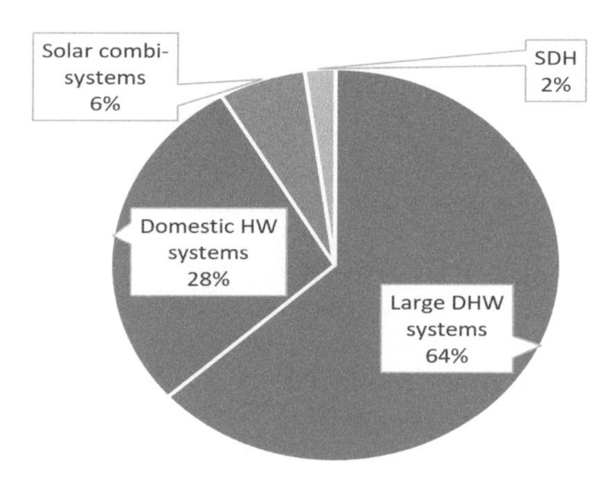

Fig. 2.9 China's solar heat consumption in 2021 (364 GW). *Data Source* IEA SHC

B. Concentrated Solar Power (CSP)

CSP development first started in the U.S. in the 1980s. The installation costs dropped quickly with CSP expansion. However, this market advancement was not sustained. When the only active company went bankrupt in 1990, CSP expansion lost its steam. A long CSP winter lingered for almost 20 years, in which no new plants were built.

Since 2006, Spain has led the global CSP market. Its leadership can be attributed to two main factors. First, the country is one of the European countries endowed with a significantly higher number of hours of sunshine than the rest of the Western European countries. Second, the Spanish CSP industry received extremely generous government support in its early market expansion.

The Spanish government's support in solar power deployment triggered the takeoff of almost 50 CSP plants in Spain. However, in contrast to the more sophisticated German feed-in tariffs (FITs), the Spanish FITs were too generous and lacked a degressive feature that would have annually reduced the support rate, added pressure on the CSP developers for technological innovation and cost reduction, and incentivized the optimal use of solar heat storage to increase the power generation efficiency and reduce CSP power generation costs.

Instead, generous government support doubled the installation cost without leading to the optimal use of solar heat storage. Although the CSP generation costs started to decline again by 2011, the Spanish FIT scheme for CSP had to be cancelled in 2013 because of the investment bubble and the government budget crisis.

Additionally, new CSP companies and engineers without experience entered the Spanish CSP market. Therefore, no price reduction took place, and the market experienced instead a sharp cost increase. Although cost reduction was able to kick in again by 2011 with the new CSP companies having gained experience, the cancelled Spanish support scheme in 2013 stopped the CSP expansion in Spain.

However, many surviving Spanish companies moved to other markets, such as Morocco or South Africa. With the spread of CSP technologies, new companies have continued to emerge worldwide, such as in the Middle East, the U.S. and China. The spread of CSP technologies allowed CSP costs to fall continuously. A recent CSP project in Australia closed at $60/MWh, which was competitive with new gas power and much cheaper than any other controllable renewable option, such as wind power equipped with battery storage.

Technologically, parabolic-trough plants overwhelmingly dominated the CSP market and had a record-high share of 90% of all CSP plants until 2010. Since then, new CSP plants have preferred solar tower technology because its operation at higher temperatures (up to 565 °C or 1049 °F) than those of parabolic troughs (up to 400 °C or 752 °F) achieved higher power generation efficiency. In 2021, the dominance of parabolic-trough plants in the global CSP market was significantly weakened. The share of global installed capacity of parabolic-trough plants larger than 50 MW decreased to 75%, whereas the share of total installed capacity of solar power tower plants larger than 50 MW exceeded 21% (Fig. 2.10).

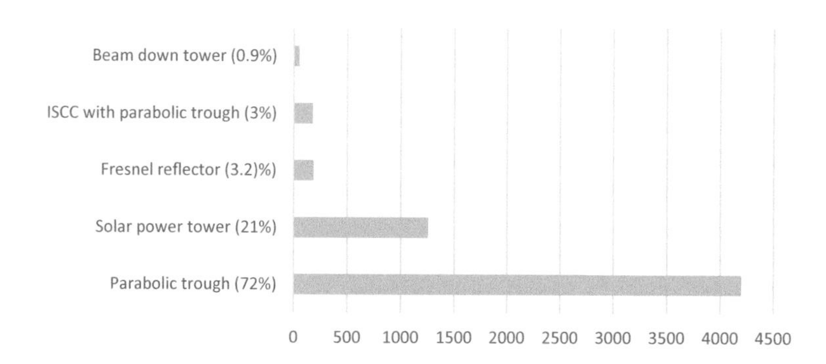

Fig. 2.10 CSP plants by technology and share in 2023. *Data Source* IRENA (2023)

The installed capacity of CSP generation grew from 419 MW in 2000 to 6.5 GW in 2022. CSP power capacity expanded the most quickly between 2008 and 2013, with a high average annual growth rate of 43%. However, because its expansion slowed significantly to an average annual growth rate of 5.4% between 2014 and 2022, global CSP power generating capacity showed a moderate average annual growth rate of 6.7% in the decade between 2013 and 2022.

In addition, although both CSP and PV use energy from solar radiation and both had similar competitive strength in the early solar power market, the subsequent deployment of CSP differed significantly from that of solar PV power. The growth of the global CSP market has been slow in contrast to that of the solar PV power market. While PV power grew more than 26 times between 2010 and 2022, CSP only grew more than four times in the same period (Fig. 2.11).

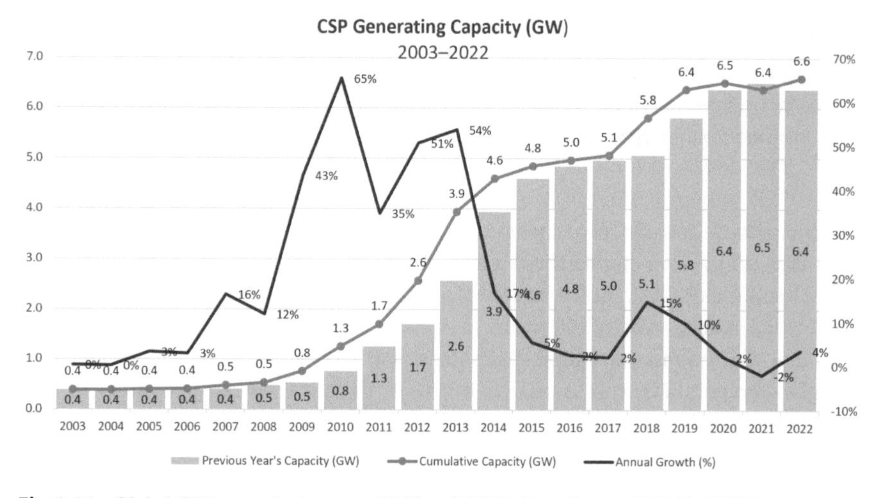

Fig. 2.11 Global CSP capacity between 2000 and 2022. *Data Source* IRENA (2023)

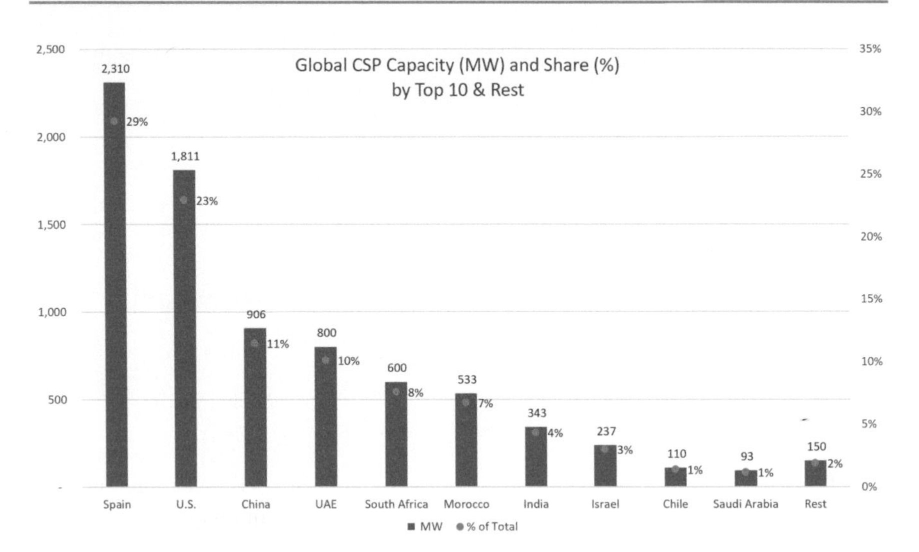

Fig. 2.12 World installed CSP capacity in 2022. *Data Source* IRENA, 2023

Currently, the total global installed CSP capacity amounts to 7.9 GW. Approximately 29% (2.3 GW) of this global CSP capacity was installed in Spain. Although the country had no new CSP capacity since 2013, it remained the world's CSP leader for more than two decades. In addition, a recent government plan announced that the country would add 5 GW CSP capacity as part of its plan to expand renewable energy to cover 74% of Spain's power demand by 2030.

Next to Spain, 23% (1.8 GW) of the world's CSP capacity was operating in the U.S. The remaining 48% (3.9 GW) was mainly contributed by China, the UAE, South Africa, Morocco, India, Israel, etc. (Fig. 2.12).

At the same time, the top U.S. CSP plants have more installed power generation capacities than those in Spain. Despite its world leadership in CSP deployment (2.2 GW), none of Spain's 27 currently operating CSP plants are among the world's five largest CSP plants. The majority of the top five CSP plants are located in the U.S. and were commissioned in 2013–2014.

The other two of the top five were larger ones and were commissioned more recently. The world's largest CSP complex, Mohammed bin Rashid Al Maktoum Solar Park, Phase 4, will have an installed capacity of 700 MW, with 517 MW currently commissioned. The second largest CSP facility is the Ouarzazate Noor Solar Power Complex in Morocco, which was commissioned in 2018. It includes two parabolic trough stations and one solar power tower and has a combined installed capacity of 510 MW.

The Ivanpah Solar Power Facility (392 MW) is the world's second largest CSP plant and uses solar power tower technology without heat storage. The remaining two CSP projects have the same installed capacity of 280 MW and use parabolic trough technology. One of them is located in California and the other in Arizona. Another important comparison is that of all three top U.S. CSP plants, only one,

Name	Country	Location	Capacity (MW)	Technology type	Storage hours	Date of Completion	Capacity factor
Mohammed bin Rashid Al Maktoum Solar Park	UAE	Dubai	700	Parabolic trough, solar tower	15	2013-2023	24.60%
Ouarzazate Noor Solar Power Station	Morocco	Ghassate, Ouarzazate	510	Parabolic trough	17.5	2018	37%
Ivanpah Solar Power Facility	U.S.	San Bernardino County, California	392	Solar power tower		2014	24.10%
Mojave Solar Project	U.S.	Barstow, California	280	Parabolic trough		2014	26.50%
Solana Generating Station	U.S.	Gila Bend, Arizona	280	Parabolic trough	6	2013	33.90%

Fig. 2.13 World's top five CSP plants

i.e., Solana Generating Station, is equipped with heat storage. However, the larger CSP complexes in UAE and Morocco are equipped with 15 and 17.5 h of heat storage, respectively (Fig. 2.13).

Because CSP generation technology is based on harnessing the sun's thermal energy for power generation, it has a major advantage over solar PV power technology. The CSP can easily store the heat energy it generates and can then use the stored heat to generate power whenever the power demand exists in the absence of sunlight. The heat storage feature overcomes the limitations determined by the intermittent nature of solar PV power. Currently, the cost for CSP with storage is approximately ¢9/kWh (same as a commercial photovoltaic system), which is poised to drop to approximately ¢5/kWh by 2030. The CSP needs to speed up its expansion development and reduce its investment cost to scale up and have a viable impact on carbon neutrality targets by 2050.

Challenges

A. Solar Heating Demand Challenges

According to the International Energy Agency, heat had a share of 50% in the global final energy consumption and was the largest global energy end-use in 2021. In the global heat market, industrial processes consume 50% of the total heat generated, space and water heating in buildings 46%, and cooking and agriculture the remainder. Because the global heat demand was mainly met by fossil fuels, the heating sector was responsible for 40% of global carbon emissions. However, solar heat has only a negligible share in the global heat market. The technological and policy barriers in the solar heating sector remain almost the same as they were a decade ago.

Direct carbon emissions from heating buildings broke a record of 2500 Mt in 2021, which represented a 5.5% annual growth from the previous year and 80% of direct carbon emissions in the building sector. Despite the growing use of efficient and low-carbon heating technologies, more than 60% of building heating demand is still met by fossil fuels. Aiming at reducing the carbon intensity of building heating by approximately 10% a year by 2030, compared with 2% a year in the last two decades, a much faster expansion of solar heat and other renewable thermal applications is urgently needed for heating and cooling in the building sector.

Although both solar PV and solar thermal (solar heat and CSP) belong to solar power, the deployment of solar thermal (especially CSP) lagged behind that of solar PV. The output of solar heat and CSP combined accounted for only 28% (i.e., solar heat 27% and CSP 1%) of the global solar energy output in 2021. There are multiple important factors that lead to inadequate solar thermal growth.

The rapidly falling cost of solar PV power, the resulting rapid increase in PV deployment and the prospects that PV will play a significant role in the future global power sector are among these factors because strong solar PV growth and penetration represent a strong competitive market force that overshadows solar thermal growth. However, the undesirable growth and penetration of solar thermal energy also has its own economic, technological, and political causes. To achieve the world's carbon neutrality objective, the growth rates and penetrations of both solar heat and CSP in the global renewable energy market must be multiplied by 2050.

B. Challenges of Solar Power in General

Although solar resources themselves are unlimited considering any conceivable energy demand on Earth, the high-level penetration of solar PV power faces many challenges in the current power grid. One of these challenges is the partial mismatch between the high demand for power and the conclusion of PV power generation around sunset. Another challenge is the lack of controllability of solar PV power generation. However, another challenge is the lack of flexibility of fossil fuel-fired power plants to reduce output and accommodate variable PV power generation.

These challenges cannot be adequately addressed without the substantial contribution of other renewable energy sources, especially dispatchable CSP. With the high penetration of solar PV power generation, the power grid's flexibility must be improved to be able to both fully utilize the variable and intermittent output from PV power generation and allow other renewable resources to extend their availability to the period when high power demand takes place, but renewable energy sources are no longer naturally available.

In 2022, CSP will account for 1% of global solar energy generation. There are several reasons for its underdevelopment. First, there are logistical and manufacturing challenges associated with CSP technology, such as its solar reflection technology depending on tracking and responding to the sun's movements.

Second, compared with modular solar PV technology, CSP technology lacks flexibility and scalability. The reflectors are giant in size and have an enormous amount of lift. They need to be supported and secured by expensive metal structures to ensure that any wind gust will not blow them away or knock them off of alignment. All these reflecting, tracking, responding, and supporting technologies raise the costs of CSP technologies.

In addition, past CSP projects also had issues of delayed and overbudget implementations and unsatisfactory performance, such as the high-profile failure of the Crescent Dunes facility in the U.S. state of Nevada, where leaks of molten salt tanks caused the CSP plant's power generation to fall to 3% of its designed capacity, which caused concerns about the economic and performance viability of large-scale CSP projects.

CSP needs accelerated expansion for important climate reasons, as we need to keep the global temperature rise below 1.5 °C according to the Paris Agreement and the energy sector integration with PV, molten salt storage, and heat pumps. According to the IEA, the global CSP is not on track with the Net Zero emissions by 2050 scenario, which requires an annual average capacity addition of 6.7 GW, which is approximately the same CSP capacity in 2021, or an annual average growth of almost 31% by 2030. In Germany and other EU countries, there is also an urgent need for green power and heat to replace the current energy demand met by the hard coal mines closed in 2018, nuclear power to be phased out by 2022, coal-fired power generation to be banned by 2030, and combustion engines to be banned by 2035.

C. CSP Challenges

1. High CSP Installation Cost

Although sunlight comes to us as a free energy source, the CSP technology used to turn it into power is currently more costly than PV technology. The significantly faster expansion of PV power than that of CSP led to a much faster cost reduction of solar PV power than that of CSP.

While the power-generating cost of PV power fell by 90% in the last decade, the global average installation cost per kilowatt for CSP dropped by almost 50% from \$9422 in 2010 to a record low of \$4746 in 2020. However, several factors, including inflation and supply chain shortages, pushed the CSP installation costs back to \$9091 per kilowatt installed in 2021, which almost erased the entire cost reduction in the last decade (see Fig. 2.14).

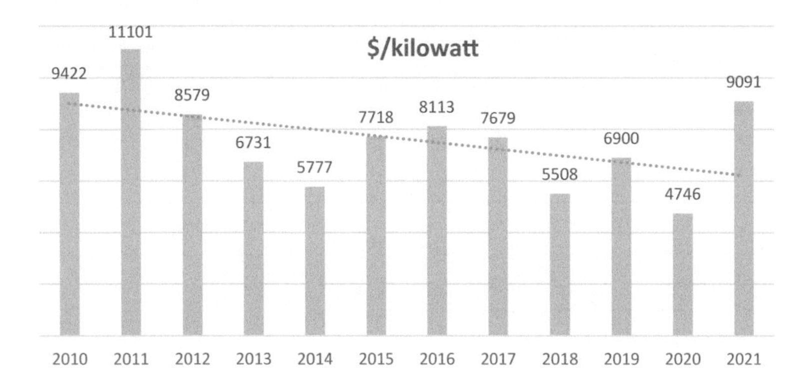

Fig. 2.14 Average global CSP installation costs. *Data Source* Statistica

2. High Land Use Challenge

Significant land use is an important barrier to CSP development in densely populated countries. In general, CSP requires more land use than fossil fuel-fired energy generation and solar PV. Several factors determine these land-use differences.

First, solar radiation has a lower energy density than conventional fossil fuel energy sources. Therefore, solar generation technologies require more land area or open space to harness solar heat and sunlight than fossil fuel-fired energy generation.

Second, like utility-scale solar PV and wind power, solar heat and CSP plants can only be most cost-effective and efficient when they are built and operate on large scales.

Third, CSP's remarkable concentration ratios from one hundred times to one thousand times are largely facilitated by larger land use than solar PV. Based on the current solar thermal energy efficiency, an average CSP plant such as a tower solar power plant, dish Stirling, or parabolic trough plant requires the use of a land area of approximately 10 acres per megawatt (MW) of power generating capacity, which is more demanding than that for solar PV power generation (6–8 acres). According to these estimates, a 200 MW power tower in the U.S. and Australia would occupy 2110 acres of land, compared to 1400 acres for a PV power plant and 940 acres for a parabolic trough plant of similar capacity (see Fig. 2.15).

These plants need considerable amounts of land for expansion. Plans to expand solar thermal plants in the most populous regions often fail not only because of higher land prices that make these solar thermal projects prohibitively expensive in these areas but also because of local concerns about such projects' competing land uses and potential visual and environmental impacts on the surrounding areas. Therefore, CSP plants are often constructed in deserts that are not suitable for residential and agricultural needs.

Technology	Capacity-weighted average land use (acres/MW)	Generation-weighted average land use (acres/GWh/y)
CSP overall	10	3.5
Parabolic trough	9.5	3.9
Tower	10	3.2
Dish Stirling	10	5.3
Linear Fresnel	4.7	4

Fig. 2.15 Average land use for CSP weighted by technology

CSP's large land use requirement for geographically small and densely populated countries such as Germany and Japan, land use for solar heat and power is an enormous barrier in many cases. Siting and permission for land use are not only extremely difficult because of the limited available land and soaring land price but also almost impossible because of solar heat and power generation's significant water use. In the U.S. and many other countries, siting and permission of land use and environmental review for solar thermal projects takes more than 3–5 years.

3. Geographic Dislocation of CSP Generation and Consumption

Similar to large-scale solar PV, both solar heat and CSP face the challenge of uneven distribution of solar resources. However, this challenge affects CSP more than solar heat. Because solar heat generation produces low temperatures, it can be deployed wherever solar PV is deployed, such as in residential areas.

In contrast, the geographic locations for CSP are even more limited in the world. CSP plants are only profitable in regions with scorching sunshine (high flux of radiant energy per unit area), such as geographic locations closer to the equator, such as Southern Europe, Northern Africa and Middle East, South Africa, parts of India, Southern U.S., Australia, and/or with consistent drier weather, such as remote unpopulated deserts in China.

The uneven solar resource distribution is an additional challenge to the further penetration of solar power when high CSP resources are located in remote unpopulated deserts, such as those in the southwest U.S. or Northwest China, while power load cities are located in solar resource-poor Northeast U.S. or Southeast China. This geographic mismatch between solar energy resources and economic centers may cause a high curtailment rate of solar power in CSP-generating regions because the generated CSP cannot be consumed locally, and the needs and costs are mounting for additional transmission systems to transmit solar power across a geographically large area across the respective country.

4. High CSP Water Use

CSP plants operating in deserts with the most abundant solar resources face another formidable challenge. Conventional CSP systems using steam turbines need a large amount of freshwater for power generation, but drought and water shortages are often major concerns in deserts where CSP plants are located. The steam turbines of CSP plants need to use freshwater as cooling water to generate the steam to be used for driving steam turbines and power generators in a process known as wet cooling. The potential cumulative impacts of CSP in regions with existing freshwater constraints raise challenging questions about whether and how to invest in the large-scale deployment of CSP.

5. Scale Requirement

While CSP plants offer a meaningful alternative to fossil fuel-fired power plants, they differ significantly from highly scalable solar PV projects. CSP plants are more demanding in terms of design, installed capacity, and output size. In contrast to a solar PV, which can be considerably scaled down to a capacity between 1 and 5 MW, the CSP's scalability is constrained by its much higher efficiency and profitability requirement of a large capacity of at least 100 MW and a large annual output of approximately 300 GWh.

6. Other Ecological Concerns

Often cited environmental and ecological impacts of solar heat and power plants include causing bird deaths, endangering wildlife habitats, inducing water pollution, and exacerbating water shortages. Therefore, CSP projects often attract public opposition and media attention, which in turn leads to increased reviews of the environmental, siting, and development risks of these projects.

7. Political Challenges

Considering the high solar heat and CSP costs, financial support from the government was essential for the existing deployment. A lack of this support is a challenge to solar heating and CSP expansion and will also negatively affect solar PV penetration. Without such government support, relatively high investment costs for solar heat, CSP, and heat storage and a lack of dedicated auctions and competition from solar PV and battery storage projects prevent faster expansion of solar heat and CSP. It is expected that less than 3 GW CSP capacity will be installed by 2026.

8. Insufficient Heat Storage Capacity

Although solar heat and CSP plants depend on expanding heat storage to improve their power generation efficiency and reduce their power generation cost, the

	# of CSP Plants	Installed Capacity (MW)	# of Plants with Storage	Share of Plants with Storage	Total Hours of Storage	Average Hours
Before 2015	38	4067	9	24%	59	6.6

Fig. 2.16 Global CSP plants (> 50 MW) with heat storage

large-scale existing plants with an installed power generation larger than 50 MW, especially those opened before 2015, are largely not equipped with heat storage.

This situation was mainly caused by the high costs of both CSP installation and heat storage, which pushed back the investment in heat storage in most of older CSP projects around the world. The majority (more than three-quarters) of CSP plants commissioned before 2015 are without heat storage. Those that are equipped with heat storage have an average storage capacity of less than seven hours (Fig. 2.16). The insufficient heat storage capacity for these old CSP plants significantly restrains their solar power generation efficiency determined by the constraints of variable and intermittent solar resources.

Solutions

A. Policy Innovations

Smart policy and technology innovations are the key to supporting the faster expansion and cost reduction of concentrated solar power and heat generation. This support can help catalyze the solar heat industry as long as the costs of its power and heat generation technologies remain too high and uncompetitive to allow it to become a major tool of the energy transition led by other renewable technologies, especially solar PV and wind power.

Policymakers can design innovative and proactive policies and regulations that delicately balance providing adequate financial incentives and keeping sufficient cost pressure. While adequate financial incentives can help solar thermal businesses grow and compete in the market, sufficient cost pressure forces solar thermal businesses to innovate and reduce costs to maintain their long-term performance competitiveness and efficiency. Innovative and proactive policies and regulations that maintain a delicate balance between government support and market competition can help the solar thermal industry quickly improve its cost competitiveness, R&D innovations, and expansion in its contribution to the world's decarbonizing energy transition.

B. Technology and Marketing Solutions

1. Upscaling the Solar Heat Market

The existing successful deployment of solar heat technologies, such as the solar district heating (SDH) in Denmark and Germany and the solar hot water (SHW) in China, provides a vital solution to the technological, economic, and environmental challenges to the further penetration of solar energy around the world.

Upscaling the SDH and SDC is a key step toward reducing the high costs of solar heating and cooling in most regions on Earth. It might be no surprise that a small thermosiphon (i.e., without using a heat pump) system (2.8 kW) with diurnal heat storage (12.7 kWh) in the Mediterranean region can supply hot water for a family of four costs for less than 2€ cent/kWh, which includes the cost of heat storage, as millions of residents already benefit from competitive solar heating solutions. At the same time, it is not easy to convince people to install such systems in cold regions because of both the higher needs of thermal energy for space heating and hot water supply and the lower amount of solar irradiation in those regions.

However, Denmark's world leadership in SDH can serve as an excellent example of what governments can do to help solarize DH systems and reduce the cost of SDH. Large SDH systems (35 MW) with seasonal heat storage (142 MWh) in small Scandinavian countries can reduce SDH costs to only €3 cent/kWh, which includes the costs of heat storage. Denmark's SDH leadership is deeply rooted in the government's proactive policy support in the form of municipalities' loan guarantees. DH companies can finance their SH projects completely with such government-guaranteed bank loans and can obtain a very low interest rate of 0–3% for a long period, such as 25 years. This financial support reduced the risk and cost of long-term investments in solarizing the country's 400 DH and CHP plants, which supply district hot water for residential heating and hot water to most of its citizens.

In addition to government financial support, high taxes on fossil fuels, and emissions trading schemes especially designed for energy distribution, the Danish success story can also be attributed to the unique features of the Danish DH utilities. These large-scale solar heating systems are mostly nonprofit entities supported by municipal governments and are both investors and operators.

2. Boosting SDH Systems with High-Temperature SH, Seasonal Storage, and Heat Pumps

To address the challenges of insufficient solar radiation and low efficiency of solar heat in the winter, leading European SDH countries, Denmark, Germany, and the Netherlands recently started to integrate high-temperature SH, seasonal storage, and heat pumps into SDH systems to boost the efficiency of the solar heat supply. Adding concentrated solar heat technologies, such as parabolic heat, with a higher-temperature (80–120 °C) solar heat supply to the SDH systems can help ensure a more stable and efficient solar heat supply in the SDH networks. Adding seasonable storage, such as low-temperature Pit thermal energy storage (a scalable and cost-efficient form of district heating energy storage), can further improve the seasonal heat efficiency, considerably reduce heat loss and supply cost by storing surplus solar heat in the summer and using it on cloudy winter days, and achieve the goal of decarbonizing DH systems. The efficiency of SDH systems can also be further enhanced by hybridizing the systems with biomass boilers or other renewable energy sources during low-sunshine months of the year. The integration of the heat pump into the existing SDH system can help reduce natural gas consumption in the winter, covering most of the heat demand of residents, businesses, and public buildings by renewable energy sources.

3. Addressing CSP Costs

A better understanding of CSP costs. Similar to the costs of other renewable energy sources, CSP costs must be compared with the costs of fossil fuel and nuclear power mainly in terms of the fixed cost of power generation. The data on future power generation technologies show that CSP installation is a more expensive option than solar PV power installation. Similar to solar PV power generation, CSP only incurs fixed operation and maintenance costs but no variable operation and maintenance cost or marginal cost, which is the cost of fuel. In addition, the environmental and health costs associated with fossil fuel combustion during power generation and consumption should not be ignored.

Further cost reduction. Innovations in CSP technologies and related policies are key solutions to allow CSP to become much more effective and cost competitive. Although the cost of CSP decreased in the past, its recent rebound needs to be adequately addressed, and further substantial cost reduction is needed to allow CSP technologies to become an affordable, expandable, and viable renewable power solution in sunny regions on Earth. Ongoing research and development in CSP technology are driving efficiency improvements and cost reductions. As the technology evolves, CSP could become more cost-competitive with solar PV, especially considering its unique advantages.

In addition, technological advancements in CSP technologies (concentrators, receiver tubes, heat transfer fluid, etc.) that lead to efficiency improvements, local manufacturing of CSP technologies, and experience and/or enhanced learning about CSP plants and their operation and maintenance can reduce CSP costs.

4. Improving CSP Efficiency with Heat Storage

Capital costs for CSP plants without heat storage are normally more than three times those for solar PV power plants. However, equipping CSP plants with heat storage can significantly improve the power generation efficiency and therefore

Technology	Capital cost (2020 $/kW)	Size (MW)	Variable operating & maintenance (2020 $/MWh)	Fixed operating & maintenance (2020 $/kW-y)	Year first available
CSP with heat storage	1,612	150	0	32.33	2022
Solar PV with tracking	1,248	150	0	15.33	2022

Fig. 2.17 Future costs of solar technologies: new CSP versus PV plants. *Data Source* NREL

significantly reduce the gap between the costs of CSP plants and those of PV power plants.

Figure 2.17 shows a comparison between the capital and fixed operating and maintenance costs for CSP plants equipped with heat storage and those costs for solar PV power plants with tracking, both with a capacity size of 150 MW. It indicates that the ratio of fixed operating and maintenance costs of the two different solar technologies to capital costs is 2.12–1, while the ratio of capital costs between the two technologies is only 1.29–1.

Heat storage provides the CSP with a desirable role in improving grid stability and flexibility in grid-integrated solar power generation. Solar heat storage technologies can store excess solar heat harnessed during sunny days, which can then be used for power generation when there is little or no sunlight, such as at night or during cloudy days. The use of heat storage can significantly improve the efficiency of CSP systems because this solar heat storing capability allows CSP technologies, in contrast to solar PV and other intermittent renewable energy, to generate power continuously day and night and regardless of the weather. Therefore, expanding CSP heat storage capacity is a major solution to the challenges of high investment costs and related efficiency of CSP, and this is especially true for large utility-scale CSP compared with large utility-scale solar PV power plants because storing thermal energy is much easier than storing power at large power levels.

Therefore, the expansion of heat storage can significantly improve CSP project efficiency and make CSP systems a far more attractive option for large-scale renewable power generation. This option can help improve the financial performance of CSP plants and the sharing capability and flexibility of solar power in the power network.

Using and adding heat storage is paramount for greatly improving CSP generation efficiency and reducing CSP generation costs in areas with the most abundant solar irradiation for power generation. Using the stored solar heat to generate power after sunset or in overcast conditions can greatly reduce the curtailment of surplus solar power and the fluctuation of the power supply caused by the surge of solar power supply when sunlight is scorching and help increase the stability, safety and efficiency of power distribution. Jointly with PV power generation, the widening price gap between cheap solar power and increasingly more expensive coal-fired power in solar-rich areas will allow CSP to play a more important grid-stabilizing role in compensating for the shortage of PV power supply in periods of sunlight absence.

Breakthroughs in heat storage technology hold great promise in solar power penetration and the fight against climate change. Improving the current and advancing next-generation heat storage technologies will allow the integration of CSP, solar PV power, and other renewable power technologies and achieve Net Zero goals in the future. The U.S. Better Energy Storage Technology Act (BEST Act) is expected to provide U.S. research initiatives to achieve innovative breakthroughs in renewable energy storage technologies aimed at reducing energy storage costs by 90% in the next decade.

The CSP output can be greatly boosted if CSP plants are equipped with high-performance heat storage. For example, adding 12 h of molten salt heat storage to a 100 MW parabolic trough plant can reduce its LCOE to less than $100 per MWh in most sunny regions, which is much cheaper than gas-fired power. At the same time, adding such molten salt heat storage to a 100 MW solar tower power plant can further reduce its LCOE to $70 per MWh, which is approximately the same as most gas-fired combined-cycle power plants and many battery-equipped solar PV or wind power plants.

Adding heat storage to CSP plants, improving the components and performance of heat storage, and enhancing the operation performance and maintenance are the necessary steps that can not only help considerably enhance the CSP plants' cost competitiveness with solar PV plants and other renewables but also significantly improve the CSP plants' performance and flexibility in power generation, their responsiveness in meeting the changing power demand, and their contribution to grid stability and safety.

The recent CSP projects using more heat storage confirmed the trend of efforts to use added heat storage as a vital solution to the CSP's challenges of high costs and weak market competitiveness.

Research shows that the CSP plants with power generation capacities larger than 50 MW installed since 2015 not only have a much higher rate (86%) of being equipped with heat storage than those installed before 2015 (24%), but the average storage capacity of the heat storage installed since 2015 has also become larger (11 h) than that installed before 2015 (6.6 h), although the average power generation capacity of the more recent CSP plants is smaller (98 GW) than that of those installed before 2015 (107 MW). Due to the above points, molten salt fluid and storage technologies are expected to see significant growth in the future solar heat and CSP market (Fig. 2.18).

5. Land and Water Use Solution

To address the challenge of CSP requiring large areas of land with high solar radiation. Using arid land unsuitable for other forms of agriculture or development to

	# of CSP Plants	Installed Capacity (MW)	# of Plants with Storage	Share of Plants with Storage	Total Hours of Storage	Average Hours
Since 2015	22	2152	19	86%	209.5	11.0

Fig. 2.18 Global CSP plants (> 50 MW) with heat storage since 2015

repurpose it for heat and power generation can reduce environmental and economic impacts in certain regions.

Some CSP technologies, such as dry cooling systems, consume less water compared to traditional power plants. This is especially valuable in water-scarce regions where minimizing water usage is a priority.

6. Improving Solar Thermal Plants

Accurately assessing solar and wind resources is vital for solar thermal power and heat generation. Solar heat and CSP plants need to use transparent, validated, and accepted performance models provided by independent third parties to accurately model the operation of the plant accounting for transient behavior of the plant, including start-ups, shut-down, intermittent clouds, and operational transitions.

Solar thermal heat and power plants need to continue improving design and equipment selection based on the experience of operating plants and further technology innovations. Solar thermal projects should hire experienced solar heat and CSP engineers with track records in appropriate engineering and equipment design, research, and development to resolve and avoid possible technology design issues for molten-salt tower and parabolic trough projects.

Because solar heat collector and heliostat technologies and the past learning process of solar heat and CSP professionals play an important role in a successful solar thermal project, it is important to select a collector/heliostat manufacture and design with a proven track record and to ensure that due quality control and proper testing and verification are successfully performed.

Solar thermal projects need more sophisticated control systems and automation of solar heat and CSP equipment. Although this will cost more, the additional investment will significantly improve the solar heat and CSP technologies' operational reliability, performance, and efficiency.

Active participation and detailed knowledge by the plant owner's team have been shown to lead to more successful solar thermal projects. The plant owner's team should prepare an appropriately detailed technical specification to be included in the EPC contract that details the key requirements and features of the plant. Solar heat and CSP plants and stakeholders need to develop solar heat and CSP standards and publicly available technical specifications for owners that reflect the lessons learned in the past. Future solar heat and CSP participants can benefit from such a standard and publicly available technical specification.

Solar heat and CSP projects should hire experienced engineers to support them during various phases. It is important that engineers play an active role in projects.

Solar thermal projects need to pay diligent attention to quality assurance/quality assessment and active owner supervision of all phases of the project work, especially on key components of equipment such as heat exchangers, turbines, and pumps.

Solar heat and CSP projects should hire experienced operation and management (O&M) companies for plant operations and maintenance. The O&M Company must be engaged and trained in time to take over the plant operation at initial acceptance and be integrated early in all phases so that engineering and procurement decisions can be informed by O&M knowledge and expertise. The O&M Company should operate the CSP plant under the supervision of the turnkey project during commissioning. CSP project teams need extensive experience and expertise in CSP technology and a track record of successful project implementation.

Summary

Solar thermal energy technologies, which include solar heat and CSP generation, are renewable energy technologies used to generate heat and power from sunlight. These technologies differ from other thermal power and heat generation technologies using carbon or nuclear energy sources because the energy source of sunlight used by these technologies is the cleanest and safest renewable energy source. Similar to solar PV power generation, solar heat and CSP generation technologies have the lowest CO_2 emissions and the lowest disease and death rates among all heat and power generation technologies. Solar thermal energy technologies have the greatest potential to offset CO_2 emissions from fossil fuel-fired heat and power generation and meet the global Net Zero target.

Despite the remarkable penetration of PV and wind power, their generation fluctuates depending on weather conditions. With higher penetrations of such fluctuating energy sources, the power grid may become unstable, and the energy supply may be unreliable. This vulnerable situation for energy supply might occur especially in the extended periods of simultaneous low wind, low sun, or overcasting conditions, which temperate regions regularly face. In those periods, a power grid only depending on PV and wind would fail even if it were backed up by battery storage.

Because solar heat collectors/plants and CSP plants can be equipped with heat storage and heat harnessed during sunny days can be stored for later use, a fleet of solar heat and CSP plants with fully controllable solar heat can also deliver heat and power at night or on cloudy days, help cover the heat and power demand in extended low sun, low wind, and overcast periods, and balance fluctuating heat and power generation. Therefore, solar heat and CSP are viable renewable energy options capable of providing utility-scale, grid-stabling, and controllable heat and power in the future of high-penetrating solar PV and wind power generation.

From a technological perspective, while solar thermal technologies differ from solar PV in their indirect power generation, they have more commonalities with geothermal technologies. First, both solar thermal and geothermal technologies deal with heat or thermal energy, which can be either used to generate power or directly used for various applications, such as hot water supply or space heating. At the same time, both solar thermal and geothermal technologies share similar challenges and need viable solutions for their further expansion and penetration.

To reduce the high investment costs of solar heat and CSP technologies, several major solutions or steps need to be taken. Examples include generating and providing funding opportunities for new technological and material scientific R&D, expanding heat storage capacity and improving storage material and efficiency, reducing red tape and related soft costs, and providing solar heat and CSP investors financial incentives or commercial loan programs to help them pay off investment over a long period.

The market conditions need to be improved for utility-scale SDH and CSP generation for selling solar heat and power and operating distributed or large-scale SD heating and cooling and CSP projects. The efficiency of these projects can be improved by adding heat storage capacity and upgrading heat storage technology. The advantages of combined solar thermal projects for both power and heat generation have become increasingly apparent. In these projects, which are close to business and residential heat consumers, the heat storage capacity of storing large amounts of high temperature heat during sunny days allows the CSP plant to generate power at night and during cloudy days and allows the solar heat plants to use surplus or waste heat for district heating and cooling, industrial and agricultural processes, etc. With added heat storage, the output and efficiency of solar heat and CSP generation can be multiplied.

Activities

Further Readings

1. U.S. Energy Information Administration (2022). Solar thermal power plants. https://www.eia.gov/energyexplained/solar/solar-thermal-power-plants.php
2. NREL (2022). Concentrating Solar Power Projects. https://solarpaces.nrel.gov/by-country/
3. (2023). Extra efficiency for Concentrated Solar Power plants via molten salt. https://helioscsp.com/extra-efficiency-for-concentrated-solar-power-plants-via-molten-salt/
4. Dieterich, R. (2018). 24-h solar energy: molten salt makes it possible, and prices are falling fast. https://insideclimatenews.org/news/16012018/csp-concentrated-solar-molten-salt-storage-24-hour-renewable-energy-crescent-dunes-nevada/

5. Epp, B. (2022). Solar heat market and industry info package 2022. https://solarthermalworld.org/news/solar-heat-market-and-policy-info-package-2022/
6. Sutu, A. (2022). Energising Europe with Solar Heat – A solar thermal roadmap for Europe. https://solarheateurope.eu/2022/05/20/energising-europe-with-solar-heat-a-solar-thermal-roadmap-for-europe/
7. Casey, J. P. (2022). High-power potential: the future of concentrated solar power. https://www.power-technology.com/features/concentrated-solar-power-hyperlight-energy/
8. Al-Kayiem, H. (2019). Solar thermal: technical challenges and solutions for power generation. https://www.researchgate.net/publication/335983852_SOLAR_THERMAL_TECHNICAL_CHALLENGES_AND_SOLUTIONS_FOR_POWER_GENERATION
9. Kraemer, S. (2022). Denmark's solar district heat leadership due to great policies. https://www.solarpaces.org/great-policy-built-denmarks-solar-district-heat-leadership/
10. Epp, B. (2022). Promising results from innovative solar district heating plants. https://solarthermalworld.org/news/promising-results-from-innovative-solar-district-heating-plants/
11. Epp, B. (2022). Policy supports gains for the Chinese solar thermal industry. https://solarthermalworld.org/news/policy-supports-gains-for-the-chinese-solar-thermal-industry/
12. O'Neil, C. & Dreves, H. (2022). Building a Solar-Powered Future. https://www.nrel.gov/news/program/2022/building-a-solar-powered-future.html

Closed Questions

1. What is the direct use of solar thermal energy?
   (a) Electric power generated from solar modules
   (b) Electric energy generated from solar heat
   (c) Solar farm
   (d) Solar tower
   (e) Space heating and cooling using solar heat collectors.
2. Which of the following is not solar thermal energy?
   (a) Electric energy generated from solar heat
   (b) Solar space heating and cooling
   (c) Concentrated solar power
   (d) Utility-scale or distributed solar PV
   (e) Solar hot water.
3. What is a glazed solar collector?
   (a) A basic component in a solar module or solar farm
   (b) A type of device that is designed for space heating or hot water
   (c) A type of electric energy storage, such as a battery

   (d) A basic component in a solar tower that converts solar heat into electrical energy
   (e) A basic component in a solar panel or installation that converts sunlight to power.
4. Which country is the world's leader in solar thermal energy?
   (a) The U.S.
   (b) Spain
   (c) China
   (d) Denmark
   (e) Germany.
5. What is the largest hurdle to concentrated solar power generation?
   (a) Low thermal conversion efficiency
   (b) Difficulty in storing the energy generated
   (c) Scorching heat
   (d) High upfront capital cost
   (e) High fuel cost.
6. What is not a challenge of solar thermal power generation?
   (a) High investment and operating costs
   (b) Lack of land for installation
   (c) Extremely hot weather
   (d) Lack of supply chain and service
   (e) Lack of water.
7. What causes the slow deployment of thermal solar power technology?
   (a) Rapid cost reduction of CSP in recent years
   (b) High competition from solar PV power generation
   (c) High rate of energy storage of solar thermal energy
   (d) High rate of solar thermal power curtailment
   (e) Low efficiency of solar thermal power generation.
8. What is not a solution to the challenges of CSP?
   (a) Mandating CSP installations in new buildings
   (b) Providing government support for research and development in CSP technologies
   (c) Further improving the efficiency of thermal collection and storage
   (d) Increasing and improving the use of molten salt in heat transfer and storage
   (e) Improving solar receiver design, heat transfer fluids, and storage materials to improve overall system efficiency and reduce costs.
9. How does concentrated solar power (CSP) work?
   (a) Solar radiation is directly converted into electricity
   (b) By using chemical reactions to capture solar energy
   (c) By converting solar energy into mechanical energy
   (d) Mirrors or lenses are utilized to focus sunlight onto a receiver to generate heat
   (e) By harnessing the energy from solar panels.

10. Which of the following belongs to solar thermal power technologies but not to concentrated solar power technologies?
    (a) Dish system
    (b) Solar updraft tower
    (c) Fresnel system
    (d) Parabolic trough system
    (e) Solar tower system.
11. How is solar thermal energy used for direct heating purposes?
    (a) Generating electricity through photovoltaic cells
    (b) Heating water for residential and commercial use
    (c) Fueling vehicles using solar-powered engines
    (d) Creating high-temperature steam for industrial processes
    (e) Cooling buildings using solar-powered air conditioning.
12. What is not one of the main benefits of solar thermal energy?
    (a) It reduces dependence on fossil fuels
    (b) It produces zero emissions throughout its lifetime
    (c) It has a low environmental impact
    (d) It can provide heating and cooling solutions
    (e) It is cost-competitive with traditional energy sources.
13. What is not one of the challenges associated with implementing solar thermal systems?
    (a) High installation and maintenance costs
    (b) Intermittent nature of solar radiation
    (c) Limited availability of suitable locations
    (d) Large land area requirements
    (e) Incompatibility with existing energy infrastructure.
14. How can the intermittency of solar thermal energy be addressed in a sustainable manner?
    (a) By incorporating energy storage systems
    (b) By using backup fossil fuel generators
    (c) By increasing the size of solar thermal installations
    (d) By reducing the use of other renewable energy sources during low solar radiation periods
    (e) By increasing the overall energy demand.
15. What is not one of the innovative solutions for improving the efficiency of solar thermal systems?
    (a) Implementing advanced tracking systems to optimize sunlight capture
    (b) Adding more reflective surfaces to solar concentrators
    (c) Incorporating nanotechnology to enhance heat absorption and transfer
    (d) Decreasing the size of solar collectors
    (e) Using hybrid solar thermal and photovoltaic systems.
16. Which of the following statements is incorrect when comparing solar thermal energy with solar photovoltaic (PV) technology?
    (a) Solar thermal is more expensive to install and maintain
    (b) Solar thermal has a higher conversion efficiency compared to PV

    (c) Solar thermal is only suitable for large-scale power generation
    (d) Solar thermal requires more land area for installations
    (e) Solar thermal has many applications beyond electricity generation.

17. Which of the following is false about the ways solar thermal and solar PV systems can be complementary?
    (a) Solar PV can generate electricity during periods of high solar radiation
    (b) Solar thermal can provide heat for PV panel cooling, improving their efficiency
    (c) PV can provide electricity during nighttime when solar thermal output is low or negligible
    (d) Solar thermal power with heat storage can offset the intermittent nature of solar PV
    (e) Solar thermal and PV can be integrated to provide both electricity and heat for a variety of applications.
18. What are negligible cost considerations associated with solar thermal energy?
    (a) Fuel costs
    (b) Higher initial capital costs compared to conventional energy sources
    (c) Ongoing maintenance and operational costs
    (d) Potential for reduced costs with technological advancements
    (e) Lifecycle costs, including replacement and decommissioning expenses.
19. Are there any environmental concerns related to solar thermal technologies?
    (a) Solar thermal technologies have no environmental impact
    (b) Solar thermal technologies generate hazardous waste
    (c) Solar thermal technologies consume large amounts of water
    (d) Solar thermal technologies contribute to air pollution
    (e) Solar thermal technologies disrupt wildlife habitats.
20. How can solar thermal energy contribute to reducing greenhouse gas emissions?
    (a) Solar thermal energy releases greenhouse gases during operation
    (b) Solar thermal energy reduces the need for fossil fuel combustion
    (c) Solar thermal energy increases the emission of ozone-depleting substances
    (d) Solar thermal energy contributes to deforestation
    (e) Solar thermal energy has no effect on greenhouse gas emissions.
21. What role does energy storage play in solar thermal systems?
    (a) Energy storage is not applicable to solar thermal systems
    (b) Energy storage allows for continuous operation during nighttime or cloudy periods
    (c) Energy storage increases the efficiency of solar thermal collectors
    (d) Energy storage helps reduce the overall cost of solar thermal installations
    (e) Energy storage negatively impacts the overall performance of solar thermal systems.
22. What are the current trends and developments in solar thermal technology?
    (a) Solar thermal technology is stagnant with no recent advancements

   (b) Current trends focus on reducing the size and complexity of solar thermal installations
   (c) Developments are focused on improving the aesthetics of solar thermal collectors
   (d) Current trends prioritize increasing the water consumption of solar thermal systems
   (e) Developments in solar thermal technology are primarily centered around cooling applications.
23. How does the location affect the viability of solar thermal projects?
   (a) The location has no impact on the viability of solar thermal projects
   (b) Solar thermal projects are only feasible in tropical regions
   (c) Ideal locations have consistent sunlight throughout the year
   (d) The location must have a high population density to justify solar thermal installations
   (e) The viability of solar thermal projects is unaffected by geographical factors.

Open Questions

1. What is solar thermal energy? What does it include?
2. What is the relationship between the temperatures, applications, and technologies of solar heat generation?
3. What is the relationship between solar thermal energy and CSP?
4. Why is the deployment of solar thermal energy paramount for combating climate change and global warming?
5. What is CSP? How does it differ from power generation using fossil fuels?
6. What types of concentrated solar power technologies are there? What are the advantages and disadvantages of these technologies? What is the range of efficiencies or conversion rates of the current CSP technologies?
7. What are the requirements for CSP? What challenges do these requirements pose?
8. How do you see the market competitiveness of CSP in comparison with that of solar PV?
9. What is the importance of higher penetration of solar heat and CSP for the entire solar energy deployment/penetration?
10. Why does the CSP have high upfront costs? What solutions are needed to help reduce the CSP price challenge?
11. What are the hardware costs and soft costs of CSP installation? Why do soft costs have a higher share than hardware costs? What options are there to reduce CSP costs?
12. What challenges does CSP power generation postpone the grid and power generation?
13. Are there any notable large-scale concentrated solar power plants?

14. What are the applications of solar thermal energy beyond electricity generation?
15. What solutions are available to address these challenges?
16. What is a utility-scale solar-heating plant?
17. What are the benefits and disadvantages of this form of solar energy application?
18. Which countries are world leaders in such solar energy projects?
19. What are the driving forces for their success stories?
20. What can the rest of the world learn from their achievements?
21. How does solar thermal energy compare to other renewable energy sources?
22. What is the future outlook for solar thermal energy and its market competitiveness?

Wind Power 3

Introduction

Science and Technology

1. From Wind Energy to Wind Power

Wind power is the use of modern renewable energy generation technology—wind turbines—to harness wind (wind energy) and convert it into usable electrical power. A wind turbine consists of propeller-like blades, rotor shafts, generators, gearboxes, and brake assemblies. The wind turns the blades of the turbine, which are connected to a generator, generating power. Multiple turbines are often grouped together in wind farms to generate a larger amount of power. The wind turbine captures the kinetic energy in the wind and converts it into mechanical power in the form of power. On a wind farm, turbines provide power to the electrical grid. These turbines can be found on land (onshore) or at sea or lake (offshore) (Fig. 3.1).

Wind turbines are manufactured in a wide range of shapes and sizes, but the most common design is the one with three blades mounted on a horizontal axis. Their output ranges from 100 kilowatts (kW) to 18 megawatts (MW). They can be installed in a wide range of locations: on hills, in open landscapes, and at sea—either fixed at the bottom of the sea or floating in deep waters.

Wind turbines typically begin generating power when the wind speed reaches the cut-in speed, which is generally approximately 5–6 m/s (or 11–13 mph). At this threshold, the turbines start turning and producing power. As the wind speed increases, the power generation of the turbines ramps up, reaching maximum capacity at approximately 12–14 m/s (or 27–31 mph/s).

To prevent damage, wind turbines are designed to shut down or enter a protective mode at high wind speeds. The cut-out wind speed, at which the turbines cease power generation, is typically set higher than 25 m/s (or 56 mph), commonly

P. Yang, *Renewable Energy*, https://doi.org/10.1007/978-3-031-49125-2_3

Fig. 3.1 Wind turbines

falling within the range of 25–30 m/s (or 56–67 mph). This safety measure helps protect the turbine components from excessive stress caused by extremely strong winds.

2. Onshore Wind Power

Onshore wind power technology involves the generation of power from wind turbines located on land. It is one of the most established and widely deployed renewable energy technologies.

Before deploying onshore wind turbines, a comprehensive assessment of wind resources is conducted. Data are collected over a period of time to determine the wind speed, direction, and consistency at the proposed site. This assessment helps identify areas with favorable wind conditions for optimal energy production.

Based on the resource assessment, suitable locations for onshore wind farms are identified. Factors such as wind speed, land availability, proximity to transmission infrastructure, environmental impact, and local regulations are considered during the site selection process.

Wind turbines are installed on land using cranes and other specialized equipment. The turbines are mounted on tower structures, which can vary in height depending on the wind conditions and turbine size. Foundations, such as concrete pads or tower footings, are constructed to provide stability and support for the turbines.

Onshore wind farms require electrical infrastructure to transmit the generated power to the grid. This includes underground or overhead electrical cables that connect the turbines to a central collection point or substation. Transformers and switchgear are used to convert and manage the power at the substation before feeding it into the grid.

The power generated by onshore wind turbines is transported to the power grid through transmission lines. These lines may require upgrades or new construction to accommodate the additional power capacity. Grid operators ensure the integration of wind power into the overall power supply and manage the balance between supply and demand.

Onshore wind farms require regular operations and maintenance activities to ensure optimal performance. This includes routine inspections, servicing of turbines and electrical components, and addressing any mechanical or electrical issues. Access to turbines is facilitated by service roads or tracks within the wind farm.

Onshore wind farms often involve engagement with local communities and stakeholders. Developers conduct consultations, address concerns, and provide information about the project's benefits, such as job creation, local economic development, and reduced greenhouse gas emissions. Community involvement and support are crucial for successful onshore wind farm deployment.

Onshore wind power technology continues to advance, with improvements in turbine design, increased hub heights, larger rotor diameters, and enhanced control systems. These advancements contribute to higher energy production, improved efficiency, and reduced costs. Scaling up onshore wind projects, including the development of wind farms with multiple turbines, helps achieve economies of scale and further cost reductions.

Onshore wind farms have environmental impacts, such as visual and noise effects, potential effects on bird and bat populations, and land use considerations. Environmental impact assessments are conducted to evaluate and mitigate these impacts. Measures such as proper siting, wildlife monitoring, and habitat conservation efforts are employed to minimize environmental effects.

The deployment of onshore wind power technology plays a crucial role in the global energy transition, providing sustainable power generation, reducing dependence on fossil fuels, and mitigating climate change.

3. Offshore Wind Power

In addition to onshore on land, wind power can operate offshore at sea or on a lake. Offshore wind power is a form of renewable energy that harnesses wind resources at sea or on a lake, complementing onshore wind farms. It offers unique opportunities for sustainable power generation with several advancements in technology and deployment.

Offshore wind farms are typically installed in relatively shallow waters, up to 50–60 m (160–200 ft) deep. These farms often use fixed foundations underwater, such as a 6 m (20 ft)-diameter monopile (single column) base, which supports wind turbines at shallow water depths of up to 30 m (100 ft). At medium water depths of 20–80 m (70–260 ft), other support structures are utilized, including conventional steel jacket structures, gravity base structures, tripod piled structures, and tripod suction caisson structures. These structures provide the necessary stability for turbines in marine environments.

Compared to onshore wind farms, offshore wind farms have several advantages, including stronger and more consistent winds, reduced visual and noise impacts, and the availability of larger and more efficient turbines.

Offshore wind farms require careful site selection to identify areas with suitable wind resources and water depths. Factors such as wind speed, water depth, seabed conditions, distance from shore, and environmental impact assessments are considered during this stage.

Offshore wind turbines are typically installed using specialized vessels equipped with cranes and jack-up legs. Foundations are constructed on the seabed, which may vary depending on the water depth and seabed conditions. Common types of foundations include monopiles (large steel tubes driven into the seabed), jackets (steel lattice structures), and floating platforms for deeper waters.

Once the turbines are installed, subsea cables are laid to connect the turbines and transmit power back to shore. These cables are buried or protected on the seabed to minimize environmental impact and potential damage from fishing activities or ship anchors.

Offshore wind farms include an offshore substation that collects and transforms the power from multiple turbines before transmitting it to the shore. The substation houses transformers, switchgear, and other electrical equipment required for power transmission.

The power generated by the offshore wind farm is transported to the onshore grid through subsea export cables. Onshore, the cables connect to an onshore substation, where the voltage is further transformed and integrated into the existing power grid.

Offshore wind farms require regular maintenance and monitoring to ensure optimal performance and longevity. Maintenance activities include routine inspections, servicing of turbines and substation equipment, and addressing any potential issues or repairs. Access to offshore sites is facilitated by specialized service vessels equipped with crew transfer capabilities.

Benefits

1. Abundant and Free Energy Source

Similar to solar power, wind power is considered one of the most established and rapidly growing sources of clean energy and the most popular green energy alternative to fossil fuels worldwide. It plays a significant role in our energy transition and carbon reduction.

Wind is caused by sunlight heating the Earth's atmosphere, the rotation of the Earth, and the Earth's surface irregularities. In other words, the difference in atmospheric pressure caused by the sun's radiation creates wind. Therefore, wind energy is an extended form of solar energy but is a motion-based form of energy. When the wind pushes turbine blades, the generator of the wind turbine converts the energy of the rotating blades into mechanical energy in the form of electric power.

Different from the fossil fuels of coal, oil, and natural gas, wind energy is free, and wind power generation does not have the additional cost of an energy source. Wind energy is also an inexhaustible energy source—as long as the Sun shines, the wind will blow. Wind blows almost everywhere, especially in remote areas. It is regularly available in the medium to long term. Considering the finite reserves of fossil fuels such as oil, gas, and coal, a large amount of wind energy for power generation is the second most important natural endowment—next to solar energy—for the energy security of humanity.

Like sunlight, wind is often available on an intermittent and variable basis. However, its variability does not follow the same pattern as that of solar energy, and both energy sources are highly complementary in terms of availability. Since large masses of energy come from air movements, there is no concern about the depletion of wind energy.

2. Environment and Health Benefits

Wind power, including onshore and offshore wind power, is a viable alternative to power generated using polluting fossil fuels because wind is a much safer and cleaner energy source than fossil fuels. Its power generation process does not emit carbon dioxide or other air pollutants. Its lifetime greenhouse gas emissions are four ton per gigawatt hour power for wind power versus 820 ton for coal, 720 ton for oil, 490 ton for natural gas, and 78–230 ton for biomass.

All emissions from the construction of a wind turbine can be offset within three to six months of operation. Then, a wind turbine can operate virtually carbon-free for the remainder of its 20-year lifetime. Comparing the carbon footprint of wind turbines (4.64 g/kWh) with that of solar panels (70 g/kWh), wind power is a more efficient renewable energy technology than solar power. In addition, the death rate from accidents and air pollution of wind power is only 0.004 deaths per terawatt hour, compared with 24.6 deaths for coal, 18.4 deaths for oil, 4.6 deaths for biomass, and 2.8 deaths for natural gas. Overall, wind power plays a vital role in combatting global warming, climate change, environmental pollution, respiratory illnesses, and lung diseases caused by CO_2 emissions, acid rain, smog, haze, and air pollution.

3. Flexible Deployment

Wind power technology can be deployed anywhere, onshore and offshore, in the world where there are abundant wind resources. More than 140,000 wind turbines are now generating power across 74 countries in the world. Many of them operate in deserts with snow or other harsh weather conditions, on hills, or at sea. This includes sites in all continents, Europe, Asia, Latin America, Africa, and Australia.

Wind turbines can be installed in a wide range of locations, including coastal areas, plains, mountainous regions, and even urban environments. This flexibility allows for the utilization of different wind regimes and the selection of sites with favorable wind resources. Wind power projects can be developed in various sizes, from small-scale installations consisting of a few turbines to large utility-scale wind farms with hundreds of turbines. This scalability enables the deployment of wind power in diverse settings, from rural communities to densely populated urban areas.

Wind turbines can be installed on land (onshore) or in bodies of water such as lakes or oceans (offshore). Although offshore wind farms are more complex to construct, they offer the advantage of stronger and more consistent winds. Flexible deployment allows for the choice of onshore or offshore installations based on the availability of suitable sites and the energy requirements of a particular region.

Wind power can be integrated with other renewable energy sources, such as solar power or hydropower, to form hybrid systems. These combined systems enhance the flexibility of renewable energy deployment by compensating for variations in generation from each source and providing a more reliable and consistent power supply. Flexible deployment of wind power involves integrating wind farms into the electrical grid infrastructure. Advanced grid management techniques, including smart grids, grid storage, and interconnections, enhance the stability and flexibility of wind power deployment. These systems help manage fluctuations in wind power generation and ensure a balanced and reliable power supply.

As wind turbine technology advances, older turbines can be replaced with newer, more efficient models in a process called repowering. Repowering existing wind farms increases their energy output and extends their operational lifespan. Retrofits involve upgrading or modifying wind turbines to improve their performance, allowing for increased flexibility in optimizing power generation.

Deployment

A. Wind Power Expansion

Wind power has experienced significant growth globally due to its cost-effectiveness, scalability, and carbon emission reduction potential. Governments, utilities, and private investors are increasingly adopting onshore wind as a key component of their renewable energy portfolios, contributing to the global transition toward clean energy sources.

In the late 1990s and early 2000s, European countries such as Germany, Spain, and Denmark were early leaders in wind power deployment. Germany has been a frontrunner in the European wind power market for many years, consistently adding substantial capacity. Spain has also made significant strides in wind power, particularly in onshore installations. Denmark, known for its early contributions to wind power, has continued to innovate and develop advanced wind turbine technologies.

Spurred by the green stimulus packages in the aftermath of the financial crisis in 2008, the U.S. wind power market witnessed remarkable growth, with several states leading the way. States with favorable wind resources, such as Texas, Iowa, and California, have witnessed substantial installations. Texas, in particular, has become a wind power giant, boasting the largest installed capacity among all U.S. states.

Similar green incentive programs have also helped China's wind power sector experience phenomenal growth since 2010, establishing itself as the largest market for wind power globally. The country's commitment to renewable energy and ambitious targets have driven rapid installations and advancements in the sector. China's robust wind power capacity has significantly contributed to global wind power capacity expansion and has played a pivotal role in driving down costs through economies of scale.

In addition to these regions, other parts of the world have also witnessed significant wind power development. India has been steadily increasing its wind power capacity, with several large-scale projects contributing to its renewable energy goals. Countries such as Brazil, Canada, and the UK have also made significant investments in wind power, further diversifying the global market. Technological advancements, such as offshore wind and floating wind, have gained traction in recent years.

With strong government support, global power generation from wind grew rapidly, mainly driven by the investment rush in China and the U.S. The two countries were responsible for 58% of the global wind power capacity in 2022. Wind power is a fast-growing renewable energy industry in the world. Over the last two decades, wind power has experienced significant growth and deployment around the world. In 2003, the global wind capacity was approximately 39 GW, but by the end of 2022, it grew to 899 GW (Fig. 3.2), which accounts for an annual average growth rate of 18.2% or a compound average annual growth rate of 17%. At the same time, it is important to see that this sharp growth of the global wind power capacity slowed down from an average annual growth rate of 24% in the first decade to 13% in the second decade.

There are several reasons for wind power's rapid development in certain countries. On the one hand, the availability of wind resources is an important factor. Countries with high wind speeds and large areas of land are particularly well suited for wind power deployment. On the other hand, rich wind resources alone would not have achieved this successful outcome without the combination of government policy support, technological advancements, and cost reductions.

Policy support, such as feed-in tariffs, renewable energy targets, and tax incentives, has played a critical role. During the late 1990s and early 2000s, several EU countries, such as Germany, Denmark, Spain, and the Netherlands, made significant investments in wind power infrastructure and implemented supportive policies. They used feed-in tariffs, tax incentives, and other mechanisms to promote the development of wind power projects. These initiatives created a favorable environment for the growth of wind power in Europe. These policies have helped to create a favorable investment environment for wind power developers and have incentivized the deployment of wind power. The success and experience gained by EU countries in wind power deployment also contributed to advancements in technology and cost reduction. These developments made wind power more competitive than traditional fossil fuel sources, further fueling its global growth.

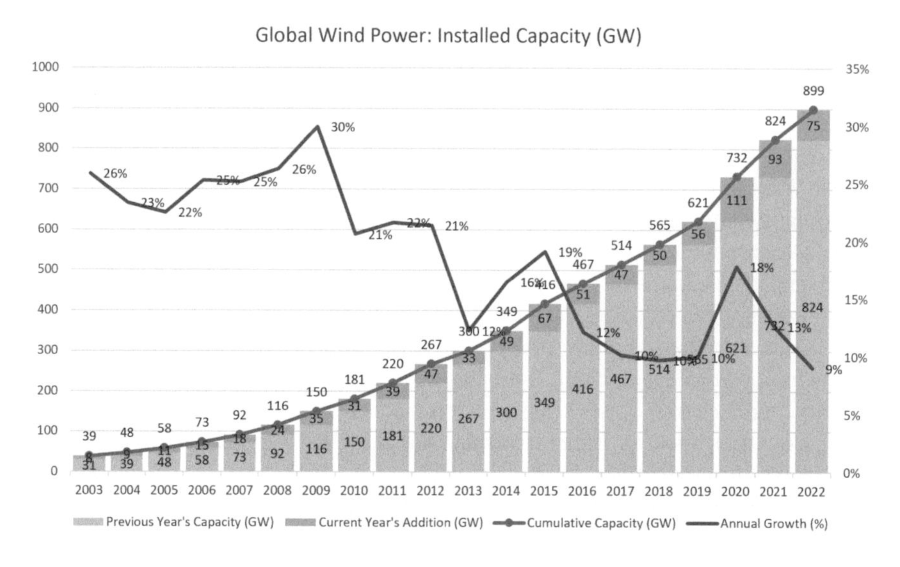

Fig. 3.2 Rapid growth in wind power capacity (2003–2022). *Source* IRENA, 2023

As a result, the growth of wind power has been particularly strong in Europe, which has been the global leader with favorable renewable energy policy in wind power deployment. Europe witnessed a rapid expansion of wind power capacity, with a substantial increase in the number of wind farms and wind turbines installed across the continent. This leadership in wind power deployment helped to drive global growth in the sector. With Europe's leadership in wind power deployment, global wind power generation started to speed up at the turn of the millennium. In 2003, Europe accounted for 73% of the global wind capacity, with a total of more than 28 GW. Denmark, Germany, Spain, and the UK were the countries that have been at the forefront of wind power deployment in Europe.

However, in the aftermath of the global financial crisis, the U.S. and China also provided generous government policy support for renewable energy deployment. The U.S. has a mix of federal and state-level policies that support wind power, including tax credits and renewable portfolio standards. Canada also has a growing wind power sector, with a focus on onshore projects.

Since then, wind power has also been growing rapidly in these countries. Since then, wind power has continued to expand worldwide, with countries from various regions, including China, the U.S., and India, becoming major players in wind power generation. Germany's world leadership in wind power was first replaced by the U.S. in 2010 and then by China. Since then, the three countries have largely maintained their world's top wind power generator positions. However, China has gradually increased its leading gap with the other two countries. In 2022, China's wind power capacity additions rebounded after a slowdown in the COVID-19 pandemic from 2018 to 2019. Wind power capacity expansion continued to accelerate most significantly in China, the U.S., and Brazil in 2022.

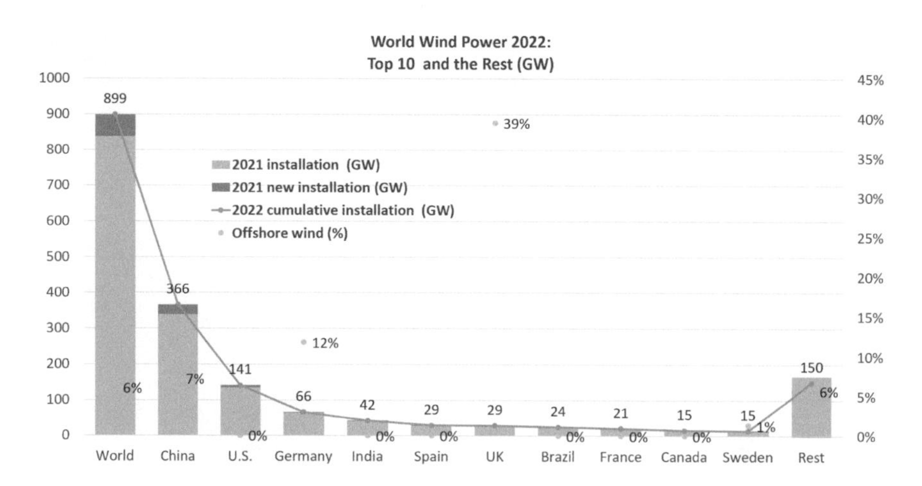

Fig. 3.3 World wind power in 2022: top 10 and the rest. *Data Source* IRENA, 2023

Wind power capacity growth slowed in India, while Sweden experienced a deployment boom. Figure 3.3 shows the new and total installed capacity of the world's top 10 wind power countries and the rest of the world in 2022. China is currently the largest market for wind power, with a total capacity of over 366 GW, accounting for almost 41% of the global wind capacity. The U.S. is now the second-largest market for wind power globally, with a substantial onshore wind capacity, particularly in states such as Texas, Iowa, and California, and a total capacity of more than 120 GW. However, Europe's early leadership and experience remain a significant factor in the growth of global wind power generation.

By region, the leadership in wind power generation shifted from Europe to Asia in 2016. Asia has experienced rapid growth in wind power deployment, driven by increasing energy demand and a focus on clean energy. The Chinese government has implemented strong policy support, including feed-in tariffs and renewable energy targets, to stimulate wind power development. India also has a significant onshore wind sector.

The gap between Asia and Europe in wind power generation has been increasing since then. Asia accounted for more than 44% of global wind power generation, and Europe generated more than half of the amount of wind power in Asia. At the time, Europe still maintained a similar share of global wind power generation to that of other regions combined. In Latin America, Brazil, Mexico, and Argentina have emerged as leaders in wind power deployment, driven by supportive policies and favorable wind resources (Fig. 3.4).

Wind power deployment in the rest of the world varies depending on regional factors. In regions such as Northern Africa and the Middle East, wind power projects are gaining traction, with countries such as Egypt, Morocco, and Saudi Arabia making significant investments. In Africa, South Africa, Kenya, and Ethiopia have notable wind power capacity. Australia has a substantial onshore

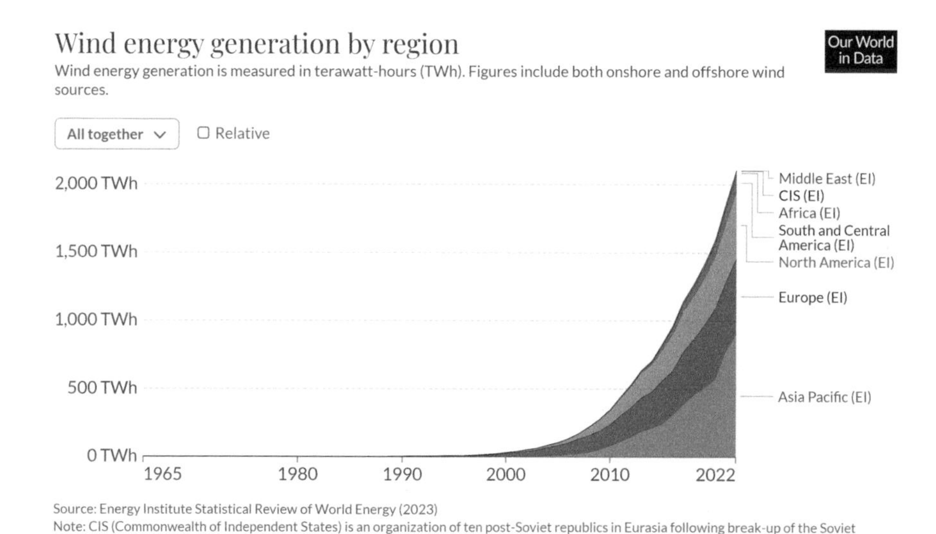

Fig. 3.4 Wind power generation

wind sector, and offshore wind projects are emerging. New Zealand is also exploring wind power as part of its renewable energy mix. Additionally, some small island nations are turning to wind power to reduce their reliance on imported fossil fuels.

Finally, technological advancements and cost reductions have also been important drivers of wind power development, making wind power increasingly competitive with traditional fossil fuel sources of power. The cost of wind power has fallen dramatically in recent years, making it increasingly competitive with traditional fossil fuel sources of power. Advances in wind turbine technology, such as larger and more efficient turbines, have also helped to improve the economics of wind power.

B. Advancements and Impacts

1. Economies of Scale

Wind turbines were extremely costly decades ago. However, as wind turbines have become much larger and more powerful over time, turbine costs have significantly declined over the past few decades due to advancements in technology, economies of scale, and increased manufacturing efficiency. In the early 1990s, wind turbine costs were relatively high compared to today. The cost per kilowatt (kW) of installed wind capacity was approximately $2000–$3000.

Since then, there has been a substantial reduction in the cost of wind turbines. The decline in costs can be attributed to various factors, including improvements in turbine design, increased hub heights and rotor diameters, better manufacturing processes, and greater market competition. Additionally, government policies and incentives have played a role in driving down costs and promoting the deployment of wind power.

As of 2021, the cost of wind turbines has continued to decrease. Onshore wind turbine costs typically range from $1000 to $2500 per installed kW, depending on factors such as turbine size, location, and project scale. Offshore wind turbines tend to be more expensive due to the additional challenges of installation and maintenance in marine environments, with costs ranging from $2500 to $5000 per installed kW.

Figure 3.5 shows the increases in the diameters in meters and the installed capacities in MW of wind turbines during 2000–2030. The approximate correlation between the increased rotor diameter of the wind turbine and the increased wind power is that the doubling of the diameter leads to the quadrupling of the installed power generation capacity.

2. Price Advantage

The dramatic price drop in the last four decades allowed wind power to demonstrate its price advantage over other energy sources. As of the end of 2021, the levelized cost of energy (total investment cost divided by the total power generated over the equipment's lifetime) of onshore wind power per megawatt-hour dropped from $330 in 1983 to $30, which made wind power the cheapest among all energy sources, including fossil fuels (Fig. 3.6).

As a result, the LCOE of wind power generation declined by 91% from 1983 to 2021 when the cumulative global deployment reached more than 824 GW to an average of $30/MWh in 2021. This price advantage trend will continue with further technological innovations of wind turbines, further economies of scale, and wind penetration (the share or percentage of wind power generation in total power generation).

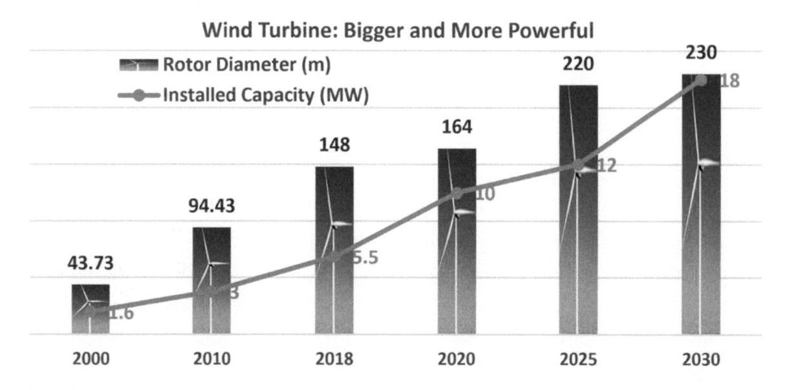

Fig. 3.5 Wind turbine size and power generation. *Data Source* Statista 2023

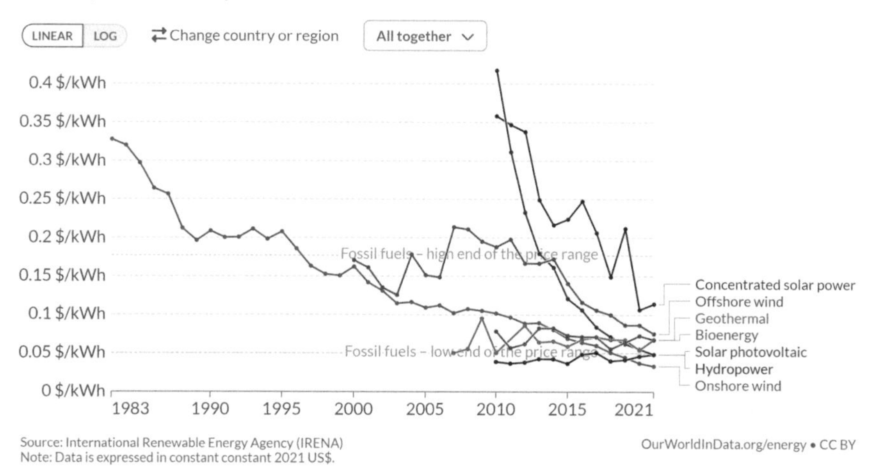

Fig. 3.6 Cost of wind power technologies versus other technologies

3. Government Support

FITs and other forms of government support around the world also incentivized large investments in wind power generation, which led to the initial wind power takeoff in Germany and established its world's wind power leadership more than twenty years ago.

Several years later, when the U.S. financial crisis in 2008 hit the U.S. economy and the world economy, massive green incentives, such as FIT, production tax credit, or investment tax credit, as part of economic rescue packages helped wind power installations experience an exponential expansion and price drop in the U.S., China, and many other countries.

This helped the U.S. and China overtake Germany as the world's top wind power generator in 2010 and 2011, respectively. Massive government support also allowed exponential growth in global wind power generation.

4. Offshore Wind Power Generation

Offshore wind power technology continues to evolve, with advancements in turbine design, floating wind farms for deeper waters, and improved installation and maintenance techniques. Governments and industry stakeholders are investing in research and development to enhance efficiency, reduce costs, and address environmental concerns.

Offshore wind farms have environmental impacts, such as potential disturbances to marine ecosystems, underwater noise, and effects on the migratory patterns of marine species. Developers undertake environmental impact assessments and work closely with regulatory bodies to minimize these impacts through careful planning, monitoring, and mitigation measures.

Offshore wind power has experienced significant growth in recent years, driven by decreasing costs, supportive policies, and the need to transition to renewable energy sources. Many countries, particularly those with favorable coastal conditions, are investing in offshore wind projects as part of their renewable energy strategies.

Offshore wind power was led by West European countries, such as the UK, Germany, the Netherlands, Denmark, and Belgium, from the beginning until 2021. Recently, China‘s wind power has started to go offshore. Its installed capacity of offshore wind power in 2022 more than tripled that in 2020, and its new installed offshore wind capacity exceeded the total offshore wind power installation of the rest of the world in the last six years. This record annual addition also allowed China’s total installation to have more than doubled that of the UK and to surpass the long-term world champion in offshore wind power in 2021 (Fig. 3.7).

The global total offshore wind power capacity exceeded 64 GW, accounting for more than 7% of the total wind power capacity. By 2025, more than 70 GW of new offshore wind power capacity is expected to be added, and the total installed fleet is projected to reach 130 GW.

The deployment of offshore wind power technology is contributing to the global transition toward clean energy sources, reducing greenhouse gas emissions, and diversifying the energy mix for a more sustainable future.

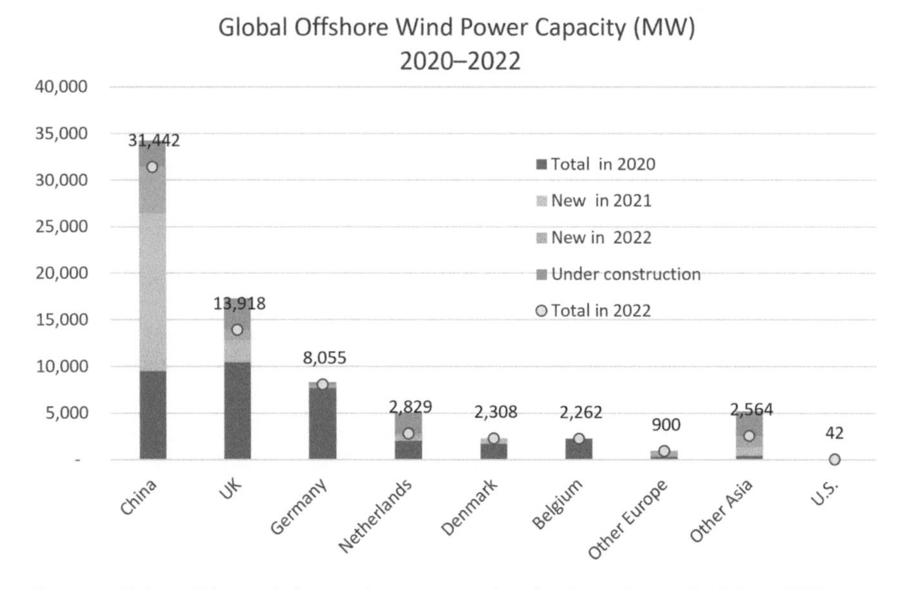

Fig. 3.7 Global offshore wind power by country and region. *Data Source* Statistica, 2023

Challenges

A. Economic Challenges

1. Urgent and Rising Green Power Demand

The urgent and rising needs for renewable energy discussed in the preceding chapter are also applicable to wind power generation. For the IEA's 'Net Zero by 2050' scenario, annual installations for wind power need to reach 160 GW by 2025 and 280 GW by 2030, and together with solar power, renewable power generation needs to be tripled and provide 90% of the total power supply by 2050.

2. Cost Surges

Price surges for many raw materials, fossil fuels, and freight costs since the beginning of 2021 caused the reversal of the long-term trend of decreasing costs of wind turbines. By March 2022, the costs of steel, copper, aluminum, and freight rose by 50%, 70%, 100%, and almost 400%, respectively, and manufacturers passed through increased equipment costs. With soaring transportation costs as the top driver of overall price surges for onshore wind, the overall investment costs of new onshore wind plants are estimated to rise by 15–25% in 2022.

3. Inadequate Wind Penetration

Despite the rapid growth of global wind power generation, the actual output of all wind power systems accounted for only 7.49% of the global power output in 2022. In more than 70% of countries, wind penetration rates were below 10%, whereas the average wind penetration rate in EU countries was 15%, and the wind penetration rates equal to or higher than 20% were limited in 10 countries, i.e., Denmark (55%), Lithuania (38), Ireland (33%), Uruguay (33%), Luxembourg (28), Portugal (28%), the UK (25%), Spain (22%), Germany (22%), and Greece (21) (Fig. 3.8).

4. High Investment and Maintenance Cost

Although wind comes to us as a free energy source, initial investment and maintenance in wind power technology are costly. On average, wind turbines cost $1.3 million per megawatt, which means that wind turbines of 2–3 MW power capacity cost approximately $2–4 million plus $42,000–$48,000 for operation and maintenance. High maintenance costs are mainly related to conventional gearbox turbines, which use a multistage gearbox between the low-speed rotor and a higher-speed generator (usually a relatively standard doubly fed induction generator). The purpose of the gearbox is to increase the rotational rotor speed (15–20 s^{-1}) to the speed required for power generation (1800 s^{-1}), commonly at a ratio of 1:90. Severe cyclic piling of random variable wind loads on the complex gearbox causes

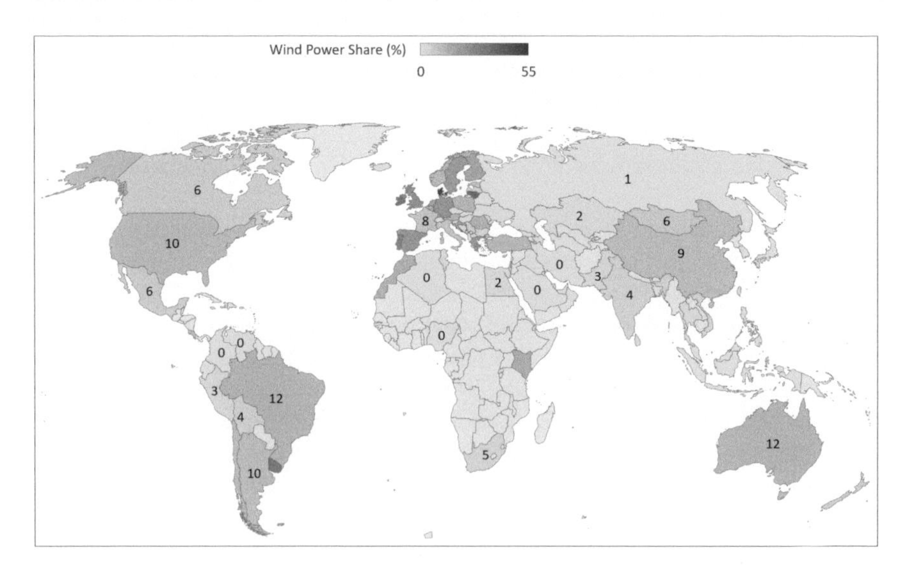

Fig. 3.8 Wind power penetration by country in 2022. *Data Source* Our World in Data, 2023

a high probability of gearbox-related failures. The maintenance and replacement of the damaged gearbox account for not only major maintenance expenses but also the highest downtime per failure of the generator (7.4 days), gearbox (6.3 days), and drivetrain (6 days) among all components of wind turbines.

5. Market Challenges

Considering the high investment cost of wind power, government support is essential for accelerating deployment. Reducing or stopping this support is a challenge to higher wind penetration. The renewable energy policy switch from FIT to auction in Germany and China has exposed further wind power energy expansion to greater market risks and challenges.

The market challenges facing wind power projects include technological lock-in, power market segregation designed for centralized power plants, market control by established fossil fuel power generators, and difficulties in overcoming their resistance, including slow interconnections and red tape of introducing wind power systems, particularly for distributed rooftop wind power generation.

B. Technological Challenges

1. High Variability

In addition to geographic location, wind power energy varies significantly depending on the geographic conditions of the site, the weather conditions, and the seasons. Wind power needs sites with abundant winds, such as the tops of smooth, rounded hills, open plains and waters, and mountain gaps that funnel and intensify wind resources, which are normally more available at higher elevations above the surface of the Earth's landmasses and oceans.

2. Geographic Dislocation of Wind Power Generation and Consumption

The distribution of wind resources is uneven. The global wind power potential map shows that a large part of Southeast China has far less wind power potential than the rest of the world. The uneven wind resource distribution can become a major challenge to the further penetration of wind power when high wind resources are located in remote unpopulated West China and power-hungry cities are located in wind resource-poor Southeast China. This geographic mismatch of wind power resources and economic centers causes a high curtailment rate of wind power in West China because of the poor local consumption of the generated power and the need and cost for an additional transmission grid to transmit wind power across China. Transmitting large amounts of variable wind power to Southeast China can add challenges and costs to wind power penetration.

3. Wind Curtailment

Wind curtailment is the forced reduction in the delivery of power generated by wind farms to the power grid. Wind curtailment takes place in wind-abundant areas and is a power regulating tool used by the grid against the variable wind power when the wind power generated is considered excessive and destructive for the grid's stability. Because it reduces the use of wind power, it is considered a waste of renewable energy. This causes a significant loss in economic and energy efficiency and adversely impacts the further penetration of wind power. Rising penetrations of variable wind power and the resulting increased variability in power systems are expected to increase wind curtailment due to oversupply or lack of system flexibility.

Wind curtailment has been a major issue in China. Until recently, the curtailment rates in some remote wind farms were as high as 50%. According to national statistics, the wind curtailment rates in China had a close relationship with the rapid development of the nation's utility-scale wind farms in remote areas, the high share of wind power generation in local power generation, long-distance grid transmission, inadequate grid transmission coordination, and high grid transmission costs. The average wind curtailment rates in China decreased from 17.1% in 2012 and 2016 to 3.0% and 3.1% in 2020 and 2021, respectively, because of the improvement in transmission grids (Fig. 3.9).

4. Land Use

Similar to solar energy, wind energy has a lower energy density than conventional energy sources and requires large wind-abundant areas to harness wind for power

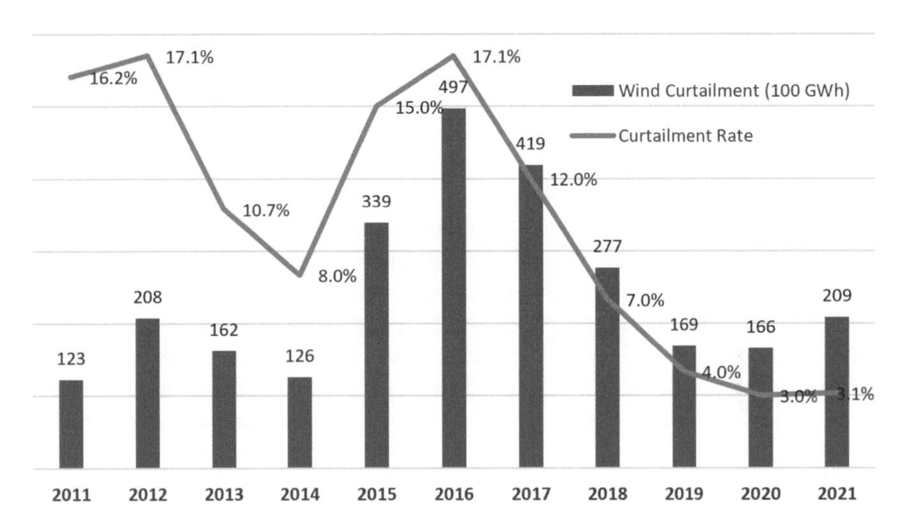

Fig. 3.9 Wind curtailment in China. *Data Source* Shi & Zhao 2018; Lou et al. 2019, NECMWC, 2022

generation. Land use is especially a major hurdle for countries with high population density and high renewable energy penetration. In Germany, for example, the reduced availability of regional development zones with the expansion of wind power causes mounting lawsuits by environmental organizations and local wind power opponents. This legal and political resistance causes, in turn, the permissions of wind power projects to become much more restrictive and lengthier, which used to take approximately ten months, but now they take up to two years.

To avoid land-use conflict with residential areas, wind turbines are often installed on agricultural or rural land, which can potentially impact farming activities and land use patterns. Farmers may need to allocate portions of their land for wind turbines, which can affect crop cultivation, livestock grazing, and other agricultural practices. Balancing land use for wind power generation and agricultural productivity is a consideration that needs to be addressed in the planning and development stages.

To optimize wind power generation efficiency, wind farms are often strategically located in areas such as deserts, where there are minimal competing land uses. However, the integration of these remote wind farms into the power grid and the transmission of the generated wind power incur additional installation and operational costs.

5. Forecasting, Control, and Monitoring

While short-term weather forecasting and wind speed prediction are relatively accurate (87–94% over 6 h and 79–85% over 24 h), it is challenging to have more accurate forecasting and predictions needed for more effective and safe wind power generation. There is also a need for using more advanced forecasting and prediction tools in combinations on different scales, such as in integration with the smart grid within the wind farm and with other wind farms. In addition, it is challenging for prospective investors to have available large-scale predictions of wind power over the typical 20-year life span of wind turbines, which climate models can help address.

Current wind power facilities still lack certainty around generation and the correlated challenges to short-term generation forecasts. This encourages the continued use of fossil fuels due to their demand reliability. By addressing the challenges that wind power currently faces, namely, operational costs, generation efficiency, and low generation predictability, smart wind technologies would dramatically improve wind power's position in the national energy mix. By being cost-competitive with the cost of fracked natural gas, these technologies will enable wind power to continue to play a central role in the ongoing sustainable energy evolution.

6. Wind and Wind Farms

The assessment of wind resources is still considered challenging for the wind generation industry because the current knowledge of the atmospheric boundary layer at heights relevant to very large turbines remains inadequate. A detailed analysis of the airflow is necessary for large wind farms with a massive number of wind turbines. Accurate assessment of airflow can help avoid interference among different types of wind turbines in large wind farms and different wind farms. Using computer resources in wind power research can simulate interference and associated loads and help develop improved control strategies to manage interference. Data-driven tools can also help locate unexpected maintenance problems, particularly with blades and drivetrains.

7. Wind Turbine Challenges

Wind power generation is considered a manure technology. Since the economics of wind power generation directly correlate with the quantity and quality of the wind to be harnessed, the size of the turbines becomes increasingly larger, and the wind blades become increasingly longer, which is driven by the enormous and expanding offshore wind market and the intrinsic complexity of turbines with blades over 100 m in length. The design of wind turbines is therefore increasingly challenging, involving an increasing number of dimensions and disciplines—aerodynamics, structure, dynamics, control, and operation.

8. Turbine Blade Challenges

Significant improvements in efficiency and reductions in cost have been achieved through the use of larger turbine blades, with the next generation of composite blade structures expected to be over 100 m in length. However, the move to ever larger blades can also create logistical barriers. Manufacturers face challenges transporting the blades to installation sites and are considering segmented blade designs that can be bonded on-site before final installation. The proposed increased size of turbine blades is limited by weight, meaning that lighter materials such as thermoplastic foams and alternative composites are being considered. Lighter blades allow for easier installation and repair as well as improved performance. However, there are inherent difficulties with composite manufacturing, such as the misalignment of fibers and inconsistent resin distribution, which can lead to lowered fatigue strength.

The impact of fatigue on turbine blades is an ongoing challenge, with each blade being subjected to more than 100 million loading cycles over the course of its lifetime. The cyclic loading of the blades is also worsened by leading-edge erosion and ice build-up.

Leading edge erosion is caused by the repeated impact of rain, ice, and particulate matter, which leads to a loss of aerodynamic efficiency and can compromise the structural integrity of the blades, leading to water ingress and UV damage. Even a small amount of leading-edge erosion can result in an ~5% drop in annual energy production.

The increased height of turbines and the span of blades both raise the risk of lightning strikes and the cost of repair. Lightning strikes can result in the loss of turbine blades and damage to electrical systems. Although lightning strike protection systems exist, failure can still occur due to moisture ingress, the detachment of diverter strips, and the erosion of blade surfaces, among other factors.

9. Inadequate Skilled Workforce and Training

The further penetration of wind power lacks the workforce with adequate scientific, technical, and manufacturing skills needed for wind power installation, maintenance, and inspection. The current educational system failed to provide adequate education and training in wind technologies necessary for wind companies to easily recruit new hires and expand their wind installation business. These issues cause inadequacies in the installation, maintenance, and inspection services of wind power systems and high soft costs.

10. Offshore Wind Challenges

The greater abundance and consistency of offshore wind attract wind power developers to exploit offshore wind in bodies of water, both sea-based and lake-based. However, offshore wind power operates in a completely different environment than onshore wind power, such as high wind, severe weather, deep water, seawater, marine wildlife, and marine ecosystems. Therefore, in addition to the challenges of onshore wind power, offshore wind power faces a set of offshore conditions, such as seawater, water depths, and severe weather, such as hurricanes, as well as protection of marine wildlife and bird migration. These conditions pose multiple technical and economic challenges.

Technically, these different ecological, environmental and weather conditions add a whole set of technical challenges for the design, manufacture, and operation of offshore wind power generation and transmission, such as corrosion, fatigue, erosion, lightning strikes, biofouling, marine pollution, and wildlife entanglement.

Economically, as offshore wind power has to address all these challenges, developing wind power offshore is currently more difficult and expensive than onshore. Adding substantially robust support structures such as foundations will increase costs by up to 30% because of the high water depth and method of construction, reduced access, and wave and weather conditions. Building transmission lines between offshore farms and onshore grids will add up to 21% of the cost of electrical infrastructure. Designing and developing advanced offshore turbines will add up to 33% of the cost. In addition, because offshore wind power entails environmental issues that impact marine wildlife and ecosystems, permitting and siting offshore wind projects are stringent and expensive. The most obvious challenges for offshore wind power are foundation challenges, which become even more severe when offshore wind farms expand to increasing water depths and operate with larger offshore turbines.

11. Hurricane Challenge

Hurricanes inflict severe damage on wind turbines and wind power infrastructure. Wind turbines, whether onshore or offshore, have built-in mechanisms that automatically stop the turbines from operating and lock and feather blades to reduce their upwind surface areas when wind speeds exceed 24.6 m/s. This places wind turbines in a "survival mode" until the storm subsides so that they can safely resume operation for power generation. Despite this automatic function of wind turbines, instantaneous changes in wind speed, wind direction, extremely turbulent flows, and lightning strikes that come with a hurricane can quickly overwhelm wind turbines' ability to resist. Hurricanes on the sea are even more devastating and destructive. In addition to the winds attacking the turbines, the turbine foundations also have to resist large, powerful waves.

12. Insufficient Storage Capacity

The FIT-supported wind power installations in Germany and other countries mainly relied on grid integration. The current insufficient energy storage capacities around the world constitute a major challenge to the further expansion of increasingly dominant intermittent wind power. Pumped hydropower storage, which is the most common, affordable, and fast-expanding form of grid power storage, has geographical limitations. Other forms of energy storage are available but are still too expensive, and their capacity is too limited to serve as grid energy storage. Energy storage capacity building is still at the initial stage. The lack of energy storage causes an ironic contradiction between the decreasing wind power costs and the increasing higher power bills for consumers and small businesses. There is an urgent need for research into completely new types of energy storage devices with potentially novel materials with physical and chemical properties.

13. Turbine Blade Disposal Challenge

Wind turbine blades require disposal or recycling when the turbines are decommissioned at the end-of-use stage or when wind farms are being upgraded in a process known as repowering. Repowering will keep the existing site and maintain or reuse the primary infrastructure for wind turbines but upgrade with larger capacity turbines. The turbine blades are replaced with more modern and larger blades. In addition, when a wind turbine reaches its operational life of approximately 25 years, approximately 85% of its components, such as copper wire, electronics, gearing, and steel, can be recycled or reused, except for turbine blades. Built to withstand hurricane-level winds, modern wind turbine blades are typically built using fiberglass, steel, sheets of balsa wood, and epoxy thermoset resin.

Therefore, spent fiberglass blades pose the greatest challenge to end-of-use considerations for wind power. The disposal of spent blades is challenging for several reasons, from the prohibitive costs of transporting them long distances because of their colossal size to finding places to store them. Dumping them in a nearby landfill would be a cost-saving solution but runs directly against the very principle of wind power as a renewable technology. It is certainly possible to cut the decommissioned blades into pieces onsite. However, this still does not solve the difficult high cost of transporting them for recycling or disposal. The cutting of extremely strong turbine blades also requires giant equipment, such as vehicle-mounted wire saws or diamond-wire saws. Because of the difficulty and limited options of recycling wind turbine blades, the vast majority of spent turbine blades are either stored in various places or dumped in landfills. Since the number of decommissioned turbine blades is bound to rise in the future, the lack of reuse and recycling is not sustainable.

14. Inadequate Transmission Grid

Transporting utility-scale wind power from remote onshore and offshore wind power farms to urban load centers where most of an economy's energy is consumed requires an adequate transmission grid. Increasingly higher penetration of intermittent wind resources and power output poses unique problems in transmission planning and efficient utilization of transmission infrastructure, resulting in higher transmission losses and costs, increased congestion, and even power curtailment with a lack of adequate transmission capacity. Due to potential transmission constraints, wind project developers will need to evaluate the economic tradeoff of sitting where the resource is best versus sitting closer to loads where transmission constraints are less likely.

Solutions

Addressing wind power challenges and maintaining the operational health of wind turbines has become increasingly important because the need for wind power is growing.

A. Technological Solutions

1. Enhancing Turbine Efficiency and Health

Despite the notable efficiency of current wind turbines, further improvement of wind turbine efficiency and health will not only help accelerate wind power penetration but also improve the public's perception of wind power efficiency and their willingness to pay for wind power.

Wind turbine efficiency improvement is an important task for scientists and engineers in wind turbine R&D. Increased efficiency can reduce the demand for wind turbines to provide the same level of power and reduce wind costs for wind turbine manufacturing, marketing, installation, and investment. Efficiency improvements will allow wind power manufacturing to use fewer resources, wind installation and generation to take up less space, and the wind power industry to reduce its environmental impacts, such as habitat loss for both plants and wildlife.

The R&D project of Professor Ken Visser at Clarkson University, which includes a novel blade design and the unique placement of the turbine rotor in the duct itself, has found a revolutionary way to harness more than 80% of the original air stream's energy, significantly more than the typical 30% of similar small turbines.

2. Improving Wind Turbine Technology

Wind turbines are becoming ever larger and more productive. This trend is driven by the enormous and expanding offshore wind farms and the complexity of the current mainstream turbines with blades over 100 m in length. Wind turbines require multidimensional improvements in aerodynamics, structure, dynamics, control, and operation. Offshore turbines also entail continued innovations in hydrodynamics, wave action, and foundation issues. Complex and sophisticated scientific and engineering tools—such as design and multiscale modeling—are utilized to examine and improve the material properties and fatigue behavior and better utilize onshore and offshore wind resources.

3. Direct Drive Wind Turbine

In addition to gearbox turbines, there are increasingly more gearless and hybrid turbines. Gearless wind turbines are called direct-drive turbines, which are a relatively new type of wind turbine. Different from conventional gearbox wind turbines, direct-drive turbines do have a gearbox. The energy of wind-pushing turbine blades directly rotates the generator without the transmission of the complex gearbox system. Instead, a magnetic or electric excitation system is installed in the direct drive wind turbine to increase the rotational speed (Fig. 3.10).

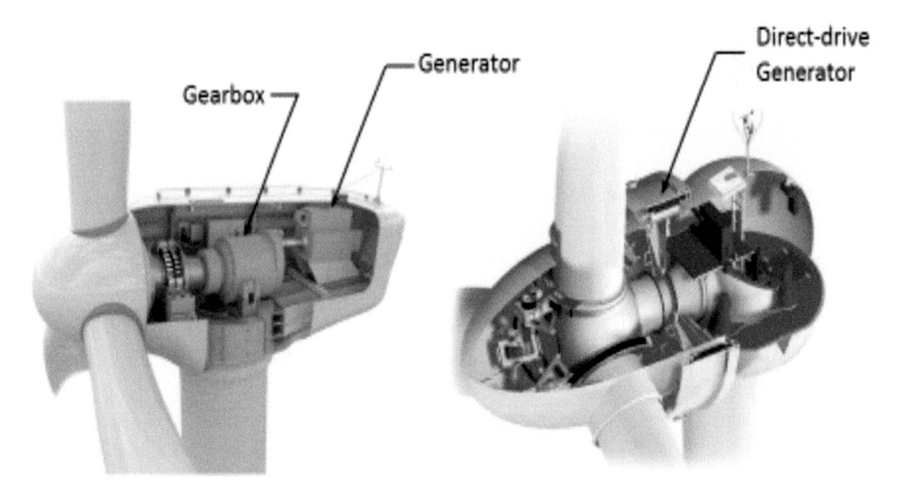

Fig. 3.10 Direct drive turbine versus gearbox turbine. *Image Credit* U.S. Department of Energy, 2019

By eliminating the gearbox, direct-drive turbines have multiple advantages. First, they can start at a very low wind speed of 1.7 m/s, a lower cut-in speed than the normal cut-in speed of 3.3–3.9 m/s for box wind turbines. Second, using permanent magnets or field coils, direct drive wind turbines significantly reduce mechanical damage to the wind turbine. The significantly reduced mechanical damage to the drivetrain of wind turbines, the major maintenance and replacement expenses and the highest downtime related to the gearbox system can be saved. Direct-drive turbines also provide more power generation reliability and efficiency and less noise.

Direct-drive turbines compete with gearbox wind turbines for recognition as the dominant design. Major wind power plant operators have begun to transition from conventional gearbox turbines to direct-drive turbines because the latter are lighter and more cost-effective than conventional wind turbines because of their reduced weight and eliminated gearbox maintenance problems and costs. Direct-drive turbines accounted for 17% of the global market in 2010, were expected to rise to approximately 24% by 2016, and had a market share of 70% of the installed new wind turbines in 2019. Currently, approximately 80–85% of wind turbines in operation have gearboxes.

The gearless design of direct-drive turbines also comes with new challenges. Since the loads from the wind turbine rotor are directly transmitted into the energy generator or motor, direct-drive turbines must use permanent magnets for excitation. These permanent magnets not only require expensive, heavy, rare earth materials such as neodymium and dysprosium but also make the generator and the converter much larger, more robust, and costly to deal with the large torque input caused by the harsh environment and keep the air gap open and stable.

4. Offshore Wind Power

In recent years, there have been remarkable advancements in offshore wind technology. Larger and more efficient turbines have been developed, enabling increased energy generation capacity. Additionally, innovative floating foundations have emerged, which allow for the deployment of offshore wind farms in deeper waters, potentially reaching depths of up to one kilometer (3300 ft) based on currently proposed technologies. These floating foundations offer promising solutions for areas with challenging seabed conditions or greater water depths.

Offshore wind power presents significant environmental benefits, including the reduction of greenhouse gas emissions, air pollution, and reliance on fossil fuels. It also contributes to the diversification of the global energy mix, enhancing energy security and promoting sustainability.

Furthermore, the costs of offshore wind power have been decreasing, driven by technological advancements, economies of scale, and supportive government policies. As a result, offshore wind is becoming increasingly economically competitive compared to traditional energy sources.

The global deployment of offshore wind power has been growing rapidly, with numerous projects being developed in various regions. However, it is important to engage with local communities, conduct stakeholder consultations, and conduct thorough environmental impact assessments to ensure the sustainable and responsible development of offshore wind projects.

Overall, offshore wind power represents a crucial component of the renewable energy transition, offering substantial energy generation potential, environmental benefits, and economic opportunities for a greener future.

Although offshore wind power requires higher costs for initial investment for installation and higher costs for operation, maintenance, transmission, and power storage, these costs can be offset in a reasonable period once offshore wind power is commercialized and gains economies of scale. This perspective is fully realistic considering that approximately 40% of the world's population lives within 100 km of the coast, and offshore wind farms will be much closer to this large coastal population than onshore wind farms in remote areas.

To address the environmental impacts of floating offshore wind, governments can develop proactive policies and laws, which can help reduce environmental and ecological impacts through avoidance, research, monitoring, and comprehensive mitigation measures (Fig. 3.11).

For example, project planning and siting processes should consider potential risks to wildlife, and siting and permitting decisions should be based on science and input from experts and stakeholders. Once potential floating offshore

Fig. 3.11 Offshore wind turbines. *Image Credit* U.S. Department of Energy

wind sites are identified, comprehensive ecological studies of these areas should be conducted, during which more data should be collected and examined on the appropriateness of a site, and multiyear baseline ecological data should be established to evaluate the impacts of an offshore wind farm to be developed, constructed and operated after permission. All offshore wind farms should be studied and monitored before, during, and after each stage of project development, construction, and operation so that ecological impacts can be continuously evaluated.

5. Smart Wind Power

Smart wind power or digitization of wind power is a vital approach to many challenges faced by wind power technology. Smart wind power covers the digitization of wind turbines and wind farms, including wind turbine and wind resource assessment and management technologies using high-fidelity computer modeling research, the physics surrounding wind turbine noise generation, integrated wind plant control, system design, system analysis, and reliability. Using digitization and AI tools to measure, record, and analyze atmospheric data, smart wind power technologies will allow wind operators to achieve enhanced power generation and more efficient material use; reduce operation, maintenance, and service costs; extend wind turbine and wind power plant life; and optimize wind power grid control and reliability.

Smart wind turbines with smart wind turbine sensing, advanced self-control, and coordinated decision-making can greatly enhance their power generation efficiency and health. Smart sensors can use precise positioning to holistically detect the turbine environment and conditions and provide effective and accurate data input according to different wind conditions and scenarios to manage and control smart wind turbines.

Advanced self-control technology uses machine-learning methods to provide perceptual information that helps turbines evaluate their environment, status, and behavior for self-adjustment, self-adaptation, and self-control. Coordinated decision-making technology enables smart wind turbines to make coordinated decisions at the station, farm, and electric system levels to handle complex future application scenarios.

Wind farms using smart wind power technologies can have wind turbines of various sizes, heights, and positions to achieve the maximum generation of wind power. The integrated design of wind power plants will allow wind farms to monitor the wind and intelligently coordinate the orientation of turbines to optimize wind exposure and achieve both optimal power generation efficiency and turbine blade lifetime. By holistically optimizing each wind turbine's power generation efficiency and lifetime in a large wind farm, smart wind power technologies will enhance wind farm efficiency, allowing more integration of wind power generation and more grid stabilization.

Engineers designing wind turbine systems use models to understand and simulate different loads of winds and waves and their impacts on wind turbines and their foundations. They need to continuously refine their models to predict turbine loading in extreme conditions.

Improving control and monitoring is also of great importance for two typical tasks of wind power generation. Enhanced control and monitoring can help maximize power output at typically low wind speeds below 12 m/s. It can also help provide a quick response to protect the turbine at higher speeds by limiting the power. Power maximizing control and monitoring can be improved using wind speed sensors such as wind lidar.[1] Accurate measurement in smart blades using devices such as vortex generators and trailing edge flaps can help safely limit power generation by pitch adjustment of the angles of the blades to avoid wind damage and extend the fatigue life of turbines. large-scale resource assessment of the atmospheric boundary layer and beyond, wind power forecasting, and integration into the power grid.

Examining and managing the intermittency of wind power is a major issue as the penetration of wind increases in many power grids. A feasible approach to achieving high penetration and grid stability is to generate accurate forecasts for wind power over a range of times depending on the need.

6. Protecting Turbines from Hurricanes

Researchers testing solutions to protect turbines from hurricanes believe that two-blade downwind turbines with improved controllers can increase offshore wind turbines' resilience to withstand hurricane damage. The yaw controller is designed to help the turbine face the correct direction, the blade pitch controller is responsible for the direction of the blades (dependent on the wind speeds), and the generator torque controller manages the amount of power to pull off the turbine and onto the grid.

7. Turbine Blade Solutions

Scientists and engineers are pursuing several approaches to tackle the challenging issues of turbine blades. One of them is finding alternate blade materials to ensure the sustainability of wind power. More research is urgently needed on alternative reusable resins and thermoplastics, along with natural reinforcements, such as bamboo and sisal.[2]

---

[1] Also known as wind monitoring lidar or laser wind sensing, wind lidar is a remote sensing device that measures wind speed and direction using optical sensing techniques.

[2] Sisal is a tropical and subtropical plant with stiff fiber that can used for rope, twine, rug, and many other purposes.

Other experiments with turbine blades are reducing wind turbine noise and modifying the designs and structures of turbines to maximize the power generation efficiency of wind turbines, such as adding a diffuser to surround the turbine rotor to induce more airflow through the blades and produce more power. Other innovations include segmented blades to reduce transportation difficulties and the costs of ever-larger turbine blades.

Because large numbers of wind turbines will be decommissioned or replaced, more creative recycling solutions for used blades are needed. Some efforts at developing alternatives are made by utilities, including two large ones in the U.S., that partner with the Tennessee company Carbon Rivers to recycle some of their spent turbine blades instead of landfilling them. One of them uses technology to breakdown and reuse fiberglass from spent turbine blades.

8. Blade Reuse, Repurposing, and Recycling

It is difficult to recycle and dispose of current turbine blades at the end-of-use stage because they are made of fiberglass reinforced by resin. However, there are increasing experiments and innovations in finding sustainable solutions.

Repurposing. When reusing or repurposing turbine blades, it is important to ensure that reuse or repurposing has the most positive impact compared to the alternative of recycling or disposing of them in landfills. A team from the Georgia Institute of Technology, University College Cork, Queen's University Belfast, City University New York, and Munster Technological University, Cork called Rewind, has analyzed and developed several innovative concepts and designs for reusing and repurposing decommissioned wind blades and helped communities design civil engineering projects such as building pedestrian bridges.

Recycling. When recycling turbine blades, it is important to ensure that the recycling process has a net positive result compared to the alternative of disposing of them in landfills. One of the recycling forms is to use the material of turbine blades. For example, LafargeHolcim, a Swiss multinational company, is exploring how to turn decommissioned wind turbine blades into sustainable building materials based on Geocycle's ongoing initiative to safely recycle and recover GE's decommissioned turbine blades after they have been removed from the turbine and shredded. This form of recycling promises 100% recycling and reductions in carbon dioxide emissions from cement coprocessing by replacing the production of cement raw materials with recycled blades. Other technologies include mechanical recycling of fiberglass or cutting blades to pellets or boards to be used in carpentry applications and recycling thermoplastic resin.

Ban on blade dumping. It is estimated that 14,000 wind turbine blades will be decommissioned over the next few years in Europe alone. To increase innovations toward reusing, repurposing, and recycling retired turbine blades, adequate market demand to incentivize building facilities for sustainable approaches to retired blades is needed. For this purpose, government policies on sustainable applications for used turbine blades are indispensable. WindEurope, an association representing the EU wind industry, called for a Europe-wide ban on landfilling turbine blades by 2025 and appealed to reuse, recycle, or recover 100% of decommissioned blades. It is working with the European Chemical Industry Council (Cefic) and the European Composites Industry Association (EuCIA) to design new projects to reuse decommissioned wind blades.

Countries with high wind power installation or penetration should consider policy mechanisms to drive the market development of alternative solutions, such as increased producer responsibilities, beyond the disposal of wind turbine blades in landfills. Governments need to support building recycling infrastructure in regions with larger portions of wind power to address decommissioned wind turbine blades.

Government Support. Government support for research and development for the recyclability of wind turbines plays an important role. The EU has been a pioneer in acknowledging and addressing the problem of wind turbine blade waste. As part of the Horizon 2020 funding program, the EU initiated several projects focused on improving the recyclability of wind turbine blades, including EcoBlade, which focused on creating recyclable wind turbine blades, and the Re-Wind project, which aimed to reuse and recycle wind turbine blades into innovative low-carbon value-added products.

Governments in several EU countries also provided similar support. The Danish government has funded, in collaboration with Vestas and Orsted, two of the world's largest wind energy companies, research into the recycling of wind turbine blades and supported the establishment of a new recycling facility specifically for wind turbine blades. The German government funded the NewWind project on sustainable methods for recycling wind turbines, with a special emphasis on the blades. The Dutch government has funded initiatives such as the Wind Turbine Recycling project to develop a 100% circular supply chain for wind turbines. The Chinese and Japanese governments have funded research projects aimed at improving the recycling rates of rare earth metals.

In the U.S., federal government entities, such as the DOE Wind Energy Technologies Office (WETO), National Renewable Energy Laboratory (NREL), and National Science Foundation (NSF), have also funded research projects on improving the sustainability and recyclability of wind turbines, particularly the composite materials used in wind turbine blades.

To help the U.S. develop a cost-effective, sustainable recycling industry for two types of materials used in wind turbines, WETO launched the Wind Turbine Materials Recycling Prize, a $5.1 million competition in 2023. The two-phase prize will incentivize competitors to present innovative technologies in the "Initiate!" phase and then demonstrate prototypes of their technologies in the "Accelerate!" phase. This funding initiative is an important step toward establishing a more sustainable and cost-effective recycling industry for wind turbine materials. It not only helps to address the environmental challenges related to wind turbine waste but also prepares the industry for future growth, as the demand for wind energy and other renewable energy sources continues to increase.

9. Wind Curtailment Solutions

Wind curtailment occurs when the power generated by wind turbines exceeds the grid's capacity to absorb it or when there is a lack of demand for the generated wind power. Wind curtailment is an undesirable situation because it results in wasted energy and reduces the economic and environmental benefits of wind power. Solutions to wind curtailment must mitigate the factors that cause wind curtailment.

Increasing Energy Storage Systems (ESS). The primary cause of wind curtailment is the surplus power from high wind power generation during low power demand periods. ESS involves the use of storage technologies such pumped hydropower storage (PHS) and batteries to capture such excess wind power. This stored energy can be released when demand is high or wind generation is low, helping to balance supply and demand. ESS can also stabilize the grid by providing ancillary services, such as frequency regulation, enhancing grid flexibility, and reducing curtailment.

Demand Response and Flexible Load Management. Demand response programs can help consumers adjust their power consumption patterns to match periods of high wind generation. This power demand adjustment might contribute to shifting energy-intensive tasks to periods in which wind power is plentiful, reducing surplus power and thus the need for its curtailment. Smart grid technologies and time-of-use pricing can be used to facilitate demand response efforts.

Advanced Forecasting and Grid Integration. Accurate wind forecasting is crucial for efficient grid integration of wind power. Improved forecasting helps grid operators anticipate periods of high wind generation, enabling them to plan and manage grid operations more effectively. By aligning grid demand with wind power availability, the need for curtailment can be reduced.

Grid Expansion and Transmission Upgrades. The lack of transmission grid capacity and transmission infrastructure is often a main cause of wind curtailment. In those cases, expanding the grid's capacity and improving transmission infrastructure allow surplus wind power to be transported to areas with higher demand or stored in regions with energy storage facilities, reducing the need for curtailment.

Market and Policy Reforms. Implementing appropriate market mechanisms, such as dynamic pricing and incentives, can encourage grid operators and consumers to adjust their behavior to match wind power fluctuations. Additionally, supportive policies that prioritize renewable energy integration and penalize wasteful practices can drive a reduction in curtailment.

Electrification of End Uses. Increasing the electrification of various sectors, such as transportation and heating, can create additional demand for wind power. Using wind power in sectors that traditionally rely on fossil fuels can help increase the overall demand for wind power and reduce wind curtailment.

Summary

Wind power is a renewable energy generation technology that generates power by mechanically converting wind energy into power. This technology differs from power generation technologies using fossil fuels because wind power is the second cleanest and safest next only to solar power; wind power generation has the lowest carbon emissions and the lowest disease and death rate among all power generation technologies. It has great potential to offset CO_2 emissions from fossil fuel-fired power generation and meet the global Net Zero target. For more than 20 years, government support in many countries has helped wind power manufacturers produce ever larger and increasingly capable wind turbines through R&D and incentivized wind farm operators to deploy wind power technologies onshore and offshore. The continued innovation and economies of scale in wind power led to wind power's competitive edge over the cheapest coal-fired power. However, wind power generation still faces many challenges, especially in offshore expansion, and solutions are urgently needed for its full penetration to meet the global Net Zero goal by 2050.

In the current wind power market, gearbox turbines still dominate, but gearless direct-drive turbines are competing for market dominance. Conventional turbines are still less expensive, but the high costs and downtime of maintenance and replacement of gearboxes significantly erode their competitiveness. The direct-drive turbines that eliminate the gearbox eliminate these costs. Although their overall costs are still high because of a much larger and more expensive generator and converter, as well as the use of a large number of permanent magnets made of expensive rare earth, direct-drive turbines may be future wind turbines because they have already incorporated smart turbine technology, which will make wind turbines more sustainable, more efficient, and more capable.

There is no single panacea that can solve the multiple challenges to wind power. To reduce the high wind power investment and maintenance cost, more proactive measures need to be taken, for example, working out stringent policies on assessing the environmental and ecological impacts of wind power generation; educating consumers on price comparison with fossil fuel-fired power in terms of avoided fuel cost and external cost; providing government funding for new wind turbine R&D and continued deployment of digitization and AI technology in smart wind turbines, smart wind farms, and smart wind power grids; expanding offshore wind power generation; reducing red tape; and improving efficiency in permitting and licensing, providing wind power investors with financial incentives or commercial loan programs.

Activities

Further Readings

1. Cole, S. (2023). 13 Compelling Wind Energy Statistics & Facts. https://theroundup.org/wind-energy-statistics/
2. Hawkins, A. & Cheung, R. (2023). China on course to hit wind and solar power target five years ahead of time. https://www.theguardian.com/world/2023/jun/29/china-wind-solar-power-global-renewable-energy-leader
3. Veers, P. et al. (2019). Grand Challenges in the Science of Wind Energy. https://www.science.org/doi/10.1126/science.aau2027
4. Wood, D. (2020). Grand Challenges in Wind Energy Research. https://doi.org/10.3389/fenrg.2020.624646.
5. Osmanbasic, E. (2020). The Future of Wind Turbines: Comparing Direct Drive and Gearbox. https://www.engineering.com/story/the-future-of-wind-turbines-comparing-direct-drive-and-gearbox
6. Willige, A. (2020). Turbines That 'Talk' to Each Other and Other Smart Tech Propulsion of the Future of Offshore Wind Power. https://www.forbes.com/sites/mitsubishiheavyindustries/2020/12/11/turbines-that-talk-to-each-other-and-other-smart-tech-propelling-the-future-of-offshore-wind-power/
7. State of Green (2017). 10 Examples of Successful Wind Energy Solutions. https://stateofgreen.com/en/news/10-examples-of-successful-wind-energy-solutions/
8. U.S. DOE Office of Energy Efficiency & Renewable Energy (2022). Carbon Rivers Makes Wind Turbine Blade Recycling and Upcycling a Reality with Support from DOE. https://www.energy.gov/eere/wind/articles/carbon-rivers-makes-wind-turbine-blade-recycling-and-upcycling-reality-support
9. Budny, R. (2023). AI: Wind energy's paragon and net-zero's protagonist. https://www.windpowerengineering.com/ai-wind-energys-paragon-and-net-zeros-protagonist/
10. Skopljak, N. (2021). World's First Floating Wind Farm Best Performer in UK. https://www.offshorewind.biz/2021/03/23/worlds-first-floating-wind-farm-best-performer-in-uk/
11. Nilsen, E. (2023). The future of wind energy in the U.S. is floating turbines as tall as 30 Rock. https://www.cnn.com/2023/05/19/us/floating-offshore-wind-energy-turbines-climate/index.html
12. Anderson, J., et al. (2023).US offshore wind power development expanding beyond the East Coast in 2023. https://www.spglobal.com/commodityinsights/en/market-insights/latest-news/energy-transition/051723-us-offshore-wind-power-development-expanding-beyond-the-east-coast-in-2023#
13. Frohböse, P. (2023). Offshore wind 2023: New ambitions! New challenges? https://www.dnv.com/article/offshore-wind-2023-new-ambitions-new-challenges--243462

Closed Questions

1. What is wind power and how does it work?
   (a) A form of solar power
   (b) Energy generated by ocean waves
   (c) Energy harnessed from the wind's kinetic motion
   (d) Energy produced by nuclear fission.
2. What are the main advantages of wind power as a renewable energy source?
   (a) High cost-effectiveness
   (b) Low impact on the environment
   (c) Inexhaustible energy source
   (d) Ability to store energy efficiently.
3. What are the main challenges or disadvantages associated with wind power?
   (a) Limited availability of wind resources
   (b) High greenhouse gas emissions
   (c) Difficulty in grid integration
   (d) Negative impact on bird populations.
4. How is wind energy converted into electrical power?
   (a) Through direct consumption by wind turbines
   (b) By converting wind into solar energy
   (c) By rotating the blades of a wind turbine to generate power
   (d) By converting wind into thermal energy.
5. What are some of the largest wind farms in the world?
   (a) Solar Valley Wind Farm, China
   (b) Tehachapi Wind Farm, United States
   (c) Gansu Wind Farm, China
   (d) Arctic Breeze Wind Farm, Antarctica.
6. How has the cost of wind power changed over the years?
   (a) It has remained relatively stable
   (b) It has decreased significantly
   (c) It has increased exponentially
   (d) It has no impact on energy prices.
7. What is the potential capacity for wind power generation globally?
   (a) Less than 1% of global power demand
   (b) Approximately 10% of global power demand
   (c) Over 20% of global power demand
   (d) Wind power can meet 100% of global power demand.
8. How does wind power compare to other renewable energy sources, such as solar power?
   (a) Wind power is more expensive than solar power
   (b) Wind power has a higher energy conversion efficiency than solar power
   (c) Wind power and solar power have similar energy production profiles
   (d) Wind power is not considered a reliable renewable energy source.

9. What are some of the environmental impacts of wind power?
   (a) Air pollution and greenhouse gas emissions
   (b) Noise pollution and visual impacts
   (c) Increased water usage and habitat destruction
   (d) Wind power has no significant environmental impacts.
10. How are offshore wind farms different from onshore wind farms, and what are their advantages and challenges?
   (a) Offshore wind farms have higher installation and transmission costs
   (b) Onshore wind farms have higher energy production potential
   (c) Offshore wind farms require less maintenance
   (d) Offshore wind farms have lower wind availability than onshore wind farms.

   1c, 2b, 3c, 4c, 5c, 6b, 7b, 8c, 9b, 10a.

Open Questions

1. What is wind power?
2. How is it harnessed for power generation?
3. How does it differ from power generation using fossil fuels?
4. What are the types of wind turbines?
5. What advantages do direct-drive turbines have?
6. What are gearbox wind turbines and direct-drive wind turbines?
7. What are the efficiencies of current wind turbines and the efficiencies of new wind turbines?
8. What is wind power curtailment?
9. What are the environmental benefits of wind power compared to traditional fossil fuel-based energy sources?
10. What are some economic benefits associated with the development of wind power projects?
11. What is offshore wind power and its potential?
12. What are the challenges faced in implementing wind power projects, both onshore and offshore?
13. What factors determine the suitability of a location for setting up a wind farm?
14. What are the advancements in wind turbine technology that have helped improve efficiency and performance?
15. What are the advantages and disadvantages of the different types of turbines?
16. What are the challenges associated with integrating wind power into existing power grids, and how can these challenges be addressed?
17. What are the main barriers to the widespread adoption of wind power, and how can they be overcome?

18. What are the potential impacts of extreme weather events on wind power infrastructure, and how can they be mitigated?
19. What are the requirements and challenges for wind power generation?
20. What role does wind power play in creating jobs and stimulating economic growth in the renewable energy sector?
21. What solutions are needed to help reduce wind power's cost challenge?
22. Why are the new wind turbines promising for the future wind power market?
23. Why does wind power installation have high upfront costs?
24. Why is wind power paramount for combating climate change and global warming?
25. How can wind power contribute to rural electrification and decentralized energy systems?
26. How do wind power installations affect local communities, and what measures can be taken to address community concerns?
27. How do wind turbines impact bird and bat populations, and what are some solutions to mitigate these effects?
28. How does wind power compare to other renewable energy sources, such as solar power, in terms of its benefits and limitations?
29. How does wind power contribute to achieving the United Nations' Sustainable Development Goals (SDGs)?
30. How does wind power contribute to energy security and diversification of the energy mix?
31. How does wind power contribute to reducing greenhouse gas emissions and combating climate change?
32. How does wind power help reduce dependence on imported fossil fuels?
33. What is AI's role in wind power deployment?

Hydropower

4

Introduction

Science and Technology

1. **Science**

Hydropower is a renewable energy technology that harnesses the energy of flowing water and converts it into electricity. It utilizes the water flowing in rivers, streams and lakes and stored in dammed reservoirs to generate power in hydropower plants. The energy in water (hydro energy) is a direct result of solar energy. Sunlight heats the Earth's surface, causing water evaporation, cloud formation, and precipitation. This water cycle moves large amounts of water between lower and higher elevations. Water energy can be considered an extended form of solar energy. Solar radiation provides the initial energy for the water cycle, which creates the energy potential in water flows. Similar to wind energy, water energy is based on motion (kinetic energy). When water pushes turbine blades or buckets, the hydro turbine's generator converts the energy from the rotating blades into mechanical energy in the form of electric power.

In summary, hydropower captures the energy in the flow of water, which is ultimately derived from solar energy. This water energy, similar to wind energy, represents motion-based kinetic energy through hydropower plants and is transformed into usable electricity.

2. Technology

(a) *Two Types of Hydraulic Turbines*

Hydropower is the most mature renewable energy technology of hydraulic turbines to convert the energy in flowing or falling waters into electric power. Hydro turbines are the crucial components of a hydropower plant that convert the kinetic

P. Yang, *Renewable Energy*, https://doi.org/10.1007/978-3-031-49125-2_4

and potential energy of water into mechanical energy, which is then transformed into electrical energy by a generator. Depending on the height of standing water (head) and water flow or the volume of water over time, two primary types of hydro turbines can be used—reaction turbines and impulse turbines.

Impulse turbines are driven by a high-speed jet of water, which is directed onto the turbine blades. This water jet's energy is used to rotate the turbine. The water does not pass through the turbine but impacts its blades, thus causing the turbine to spin. After the water hits the blades, it falls to the bottom of the turbine and can be discharged back into the river.

The Pelton wheel is the most common type of impulse turbine. It is best suited to high head, low flow applications. The water jet, ejected from a nozzle, strikes the double-cupped buckets attached to the wheel's rim, causing it to rotate.

Unlike impulse turbines, reaction turbines are completely submerged in water and are turned by the pressure of water as it passes through the blades of the turbine. The pressure drop occurs across both the turbine blades and the fixed diffuser blades (stator). In these turbines, the water fills the runner passages, and as the water flows through the blades, it causes a reaction force that results in the rotation of the blades.

The two most common types of reaction turbines are Francis turbines and Kaplan turbines. Francis turbines are used for medium head, medium flow applications. They have a radial inflow design, with water entering the runner and flowing outward.

Kaplan turbines are used for low head, high flow applications. They have a propeller-type design, with adjustable blades that can be optimized for efficiency at varying water flow rates.

These two types of turbines, impulse and reaction, are designed to cater to different water flow and head conditions, making it possible to generate hydroelectric power in a wide variety of geographic and climatic conditions. The appropriate type of turbine is chosen based on the specific characteristics of each hydropower site. Figure 4.1 compares the differences between the two types of turbines in science, technology, efficiency, and models.

(b) *Three Types of Hydropower Pants*

There are three different types of hydropower plants: impoundment, diversion, and pumped storage. The first two types differ from each other in whether a dammed reservoir is used to store water flowing from upstream of the river before it is used to generate power. The third type differs from the first two in its primary function of serving as energy storage instead of pure power generation.

An impoundment hydropower plant is one of the most common types of hydropower plants. This type of hydropower plant is a large hydropower system with a reservoir and dam. Reservoir and dam hydropower, also known as hydroelectric power, is a method of generating electricity by utilizing the kinetic energy of flowing or falling water. This renewable energy source has been widely used for decades and continues to play a significant role in the global energy mix.

	Reaction Turbines	Impulse Turbines
Science	The stream passes through a pressure casement around the rotor before passing through the rotating blades.	The stream passes via the nozzle which produces a water jet.
Technology	Using the combined forces of pressure and moving water to rotate blades.	Using a high-speed water jet or stream to hit each of the buckets on the turbine runner
Application	Used for lower head and large flow systems	Used for relatively high and medium water heads and low flows
Efficiency	Higher efficiency because the combined forces of pressure and motion	Lower efficiency
Models	Francis turbine, Kaplan turbine, kinetic turbine, propeller turbine, bulb turbine, tube turbine	Pelton turbine, cross-flow turbine
Lowest design flow	Francis turbine: 40% of design flow Kaplan turbine: 20%-40% of design flow	1—20% (depending on the number of nozzles)

Fig. 4.1 Types of hydro turbines

Dams for this type of hydropower plant require an artificial head pond or reservoir to store water and raise the water level on one side of the dam. Water from the reservoir flows through a turbine, spinning it, which in turn operates a generator to generate power (Fig. 4.2).

A dam is built across a river or a waterway to create an artificial reservoir or a large, elevated body of water. The dam is usually constructed in a location where there is a significant difference in water level between two points, allowing the water to flow from a higher elevation to a lower elevation.

When the dam is completed, water is impounded, and a reservoir is formed behind it. The reservoir serves as a storage facility for water, and its depth depends on factors such as the size of the dam and the volume of water available.

Near the base of the dam, an intake structure is installed, which is an opening that allows water to be drawn from the reservoir. Connected to the intake is a penstock, a large pipe or tunnel through which the water flows from the reservoir to the power station.

At the power station, the flowing water is directed through turbines. These turbines have blades that capture the energy from the moving water and convert it into mechanical energy. The turbine spins a shaft connected to a generator.

As the generator's shaft rotates, it produces electricity by inducing a flow of electrons in coils of wire within the generator. This process generates alternating current (AC) electricity.

The generated electricity is then transmitted through power lines to homes, businesses, and industries for consumption.

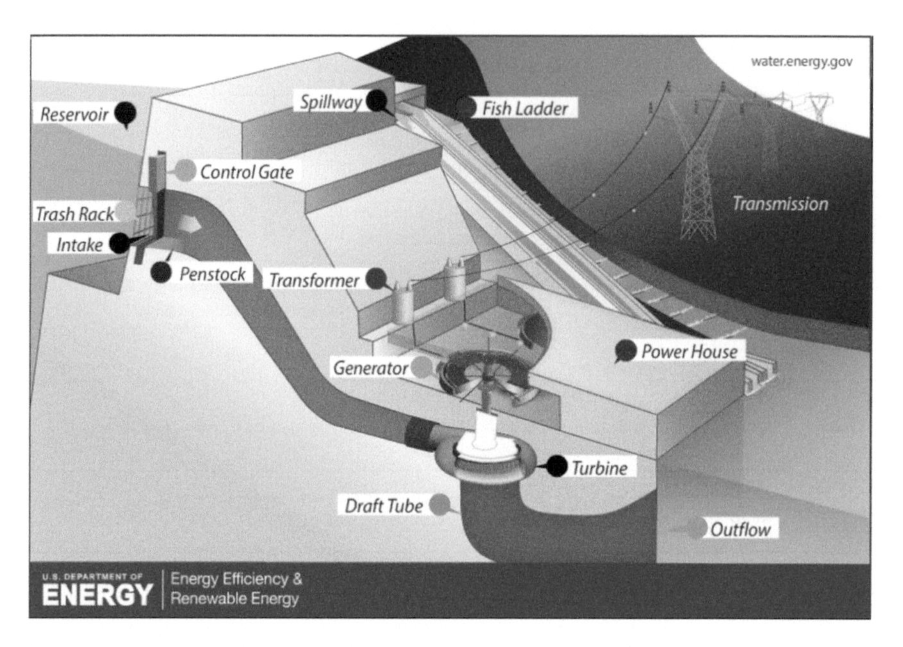

Fig. 4.2 Reservoir and dam hydropower plant

The main advantage of impoundment hydropower is its ability to provide reliable and consistent power generation. By controlling the flow of water, hydroelectric power plants can quickly respond to changes in electricity demand, making them valuable assets for grid stability.

A diversion hydropower plant, also known as a run-of-the-river hydropower plant, is a form of renewable energy system that optimally harnesses hydro energy from the natural downward flow of rivers. Typically, it is a small hydropower setup without a traditional dam and reservoir for water storage, significantly reducing environmental impact (Fig. 4.3).

This technology works best in regions with fast-moving rivers exhibiting steady seasonal water flow, making the use of a dam optional. Water is collected from a high point in the river and diverted into a pressurized pipeline or a channel, also known as a penstock. If the river possesses a steep grade or sufficient water flow conducive for power generation, the dam employed is classified as a run-of-the-river dam. This dam variant does not obstruct the entire river, allowing for seamless fish passage.

The diverted water, channeled into a canal or piped to a generating station, cascades downhill and passes a turbine, instigating a spin. The mechanical energy from this spinning turbine is converted into electrical energy by a connected generator. Once the water navigates through the turbine, it returns to the river, maintaining a minimal impact on the surrounding environment. The balance of ecosystem preservation and power production renders run-of-the-river hydro systems popular in areas with consistent annual water flow.

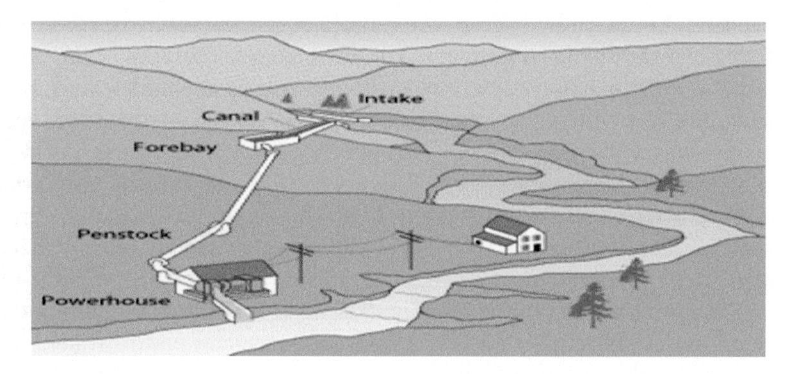

Fig. 4.3 Run-of-the-river power plant. *Image Credit* U.S. Department of Energy

Unlike their intermittent renewable energy counterparts, wind and solar power, diversion hydropower systems can produce power continuously. However, the amount of power they generate is limited by the river's volume and flow, which can fluctuate due to seasonal variations or other environmental factors.

Pumped hydropower storage (PHS) is a different use of hydropower technology. It is not intended as a facility for power generation but as a giant storage of variable renewable energy, such as wind power. Technically, a pure PHS facility includes two reservoirs, one at a lower elevation and another at a higher elevation. When wind power or solar power is in oversupply during the off-peak hours, a PHS facility uses the excess power to pump water from the lower reservoir to a higher reservoir. When power is in high demand during peak load hours, the PHS facility "discharges the stored power" to the grid by releasing the stored water from the higher reservoir back to the lower reservoir. Because PHS does not add power to the grid, the installed capacity of pure PHS is normally not included in the total hydropower generation capacity. This chapter will provide a brief introduction to PHS and its expansion as one of the solutions to the challenges of hydropower technology. Its full coverage will take place in Chap. 7, "Energy Storage."

Benefits

1. Abundant, Free and Renewable Energy Sources

Like solar and wind power, hydropower is a renewable source of energy. This technology differs from power generation technologies using fossil fuels because hydropower uses free river water instead of fossil fuels to generate power. Like solar power, hydropower is the safest energy source, safer than wind power, and much safer than all fossil fuels. At the same time, although it is not as clean as solar and wind power, it can offset most CO_2 emissions from fossil fuel-fired power generation, nearly 96% of coal, 95% of oil, or 93% of natural gas. Therefore, it can play a vital role in our climate action to meet the global Net Zero target.

The energy harnessed by hydropower relies on the water cycle, which is driven and made renewable by solar radiation. Different from fossil fuels—coal, oil, and natural gas—water energy is free, and hydropower generation does not have the additional cost of an energy source. Water energy is also an inexhaustible energy source—as long as the sun shines, water will evaporate from rivers, lakes, and oceans, condense in the atmosphere, and return to the surface of the Earth as precipitation, which will then flow from high altitudes to low altitudes through rivers and streams into lakes and seas. This moving water contains energy that can be harnessed to generate power by rotating turbines. A typical hydropower plant channels the flowing water through a pipe (known as a penstock) to turbines, and the water forces turbines to rotate a generator that generates power. Water energy can be seen in action almost everywhere, especially in the summer months. Considering the finite reserves of fossil fuels such as oil, gas, and coal, a large amount of water energy for power generation is an important natural endowment for the sustainable energy security of humanity. Because of the seasonal and geographic variation in its availability, water energy is also considered an intermittent and variable energy source. However, compared with sunlight and wind, water energy is more consistently available as a renewable energy source.

2. Environment and Health Benefits

Its power generation process does not emit carbon dioxide or other air pollutants. Its lifetime greenhouse gas emissions are 34 ton per GWh of hydropower, which means that hydropower would offset approximately 93–96% of the greenhouse gases of fossil fuel-fired power. All emissions from the construction of a hydro turbine can be offset between 2 and 4 years of operation. Then, a hydropower plant can operate virtually carbon-free for the remainder of its 20-year lifetime. In addition, the death rate from accidents and air pollution of hydropower is only 0.02 deaths per terawatt hour, making it 1230 times safer than coal-fired power, 920 times safer than oil-fired power, and 140 times safer than oil-natural gas-fired power. Therefore, hydropower plays an important role in combatting global warming, climate change, environmental pollution, respiratory illnesses, and lung diseases caused by CO_2 emissions, acid rain, smog, haze, and air pollution. Increased hydropower will lead to reduced environmental and health costs, better overall environment and air quality, and better health of citizens and workers.

3. Price Advantage

As the oldest and most mature renewable energy technology, hydropower has a low-price edge in the renewable energy market. This price advantage makes hydropower an increasingly valuable component of the energy transition. This is especially true in the developing world, where financial resources are limited. The levelized costs for hydropower per MWh vary by country and by the size of hydropower capacity.

	High - New dam	Average LCOE-New Dam /MWh	Low LCOE-New Dam /MWh	Low - Current Dam	China LCOE /MWh	Brazil LCOE /MWh	India LCOE /MWh	North America LCOE /MWh	Europe LCOE /MWh
Large (>10 MW)		$50	$20		$40	$40		$80	$120
Small (1-10 MW)					$40	$60	$60	$64	$130
Total Costs	$4,500,000/ MW		$600,000 / MW	$450,000 / MW					

Fig. 4.4 Hydropower costs in selected countries

The LCOEs of hydropower also vary depending on the scales of the projects, and smaller hydropower plants cost more for power generation. Figure 4.4 shows that costs are lowest in China, higher in Brazil, India, and North America, and highest in Europe. In the U.S., hydropower costs $64 per MW to generate compared to $400 for oil-fired power.

In addition to power generation, hydropower facilities with reservoirs are also the cheapest way to store large amounts of energy, achieve power grid stability and provide additional grid services. Therefore, PHS also plays an extremely important role in achieving the high penetration of variable renewables and the unsurpassed ability to provide grid flexibility.

4. Government Support

Government direct investments or contracts, FITs, power purchase agreements (PPAs), and other forms of government support around the world were indispensable for large investments in hydropower generation. Government contracts led to the initial hydropower takeoff in the U.S. and established its early world hydropower leadership by the end of the nineteenth century, after the Great Depression, and during and after World War II.

In 1990, the German Renewable Energy Feed-in Law offered FIT at 65% of the residential power price to hydropower projects smaller than 5 MW. In the aftermath of the U.S. financial crisis in 2008, the U.S., China, and more than 50 other countries introduced green incentives, such as production tax credit or investment tax credit (FIT), as part of the economic rescue package. Most countries capped their incentives for small hydropower projects. The U.S. provided a production tax credit or investment tax credit at 30% of the production or investment costs of renewable energy projects.

While the FITs for renewable energy technologies were largely determined by bidding or tendering, the FITs for hydropower plants, including pumped storage power plants, were set in a more complicated and individual manner. In general, hydropower plants supplying power in a province received standard benchmark tariffs or seasonally differentiated tariffs in water-abundant regions, whereas hydropower projects for interprovincial and cross-regional transmission received negotiated tariffs. Market-determined hydropower incentives significantly boosted hydropower, especially in China.

5. Economic and Financial Benefits

In addition to displacing carbon emissions by fossil fuels, the cheap, clean, and reliable energy generated by hydropower also delivers additional economic and financial benefits to the local and national economy. It reduces the shortage of power supply and increases local and national income by providing electricity. It will help reduce the reliance on imported fossil fuels, reduce the need for foreign currency, and reduce exposure to financial risk for the national economy. Increased hydropower penetration will help support rural economic development with increased irrigation, water supply, flood control, water-based transport, and tax revenues for the government. Hydropower has many advantages over fossil fuels such as coal, oil, and natural gas. Water harnessed to generate power can be reused downstream for irrigation and other purposes once it has passed through the hydropower plant.

6. Flexible Power

Different from renewable energy sources such as solar and wind, a hydropower plant with a dam and reservoir generates reliable power, which is considered baseload power. Additionally, because hydropower plants can start easily and have regular power output rapidly, they can also be used as "peak load power" suited to meet the peak demand. Therefore, increasing the share of hydropower can provide both robust renewable baseload power that is needed to displace fossil fuels and added renewable peak load power to balance and stabilize the power supply.

Deployment

A. **History**

Humans have been using energy in water to perform several tasks for thousands of years. Water energy was used for grounding grains, lifting water, irrigating, and shipping using waterwheels, water mills, and sails. Water-driven machines played a crucial role in the early stages of industrialization, especially in the textile industry in eighteenth-century Britain.

Shortly after Thomas Edison invented electric power over 100 years ago, the kinetic energy of the flowing or falling water in rivers, streams, and lakes started also being used for electric power generation. William Armstrong invented the world's first hydropower system in England in 1878 to power a single arc lamp in his art gallery. In the following decade, the use of hydropower began to increase rapidly during the industrial revolution. The first hydroelectric power plant was built at Niagara Falls in 1895.

In the early twentieth century, several developed countries, particularly the U.S. and Europe, started harnessing hydropower for power production. Commercial companies built many more small hydropower plants in mountains near cities.

Two hundred hydropower plants were constructed in the U.S. alone. By 1920, hydropower accounted for 40% of the power generated in the U.S.

In the 1930s, hydropower plants became larger, and their associated dams took on additional tasks, such as irrigation, flood control, and navigation. Large-scale hydropower plants were constructed, contributing significantly to the national power grids.

The Hoover Dam hydropower plant was constructed between 1931 and 1936 during the Great Depression and became the world's largest hydropower plant, with an initial installed capacity of 1.3 GW in 1936. Its ranking was subsequently overtaken by many much larger ones, including first the Grand Coulee Dam hydropower plant in 1942 at 6.8 GW. With water stored using dams and reservoirs, modern hydroelectric turbines can generate power in a more stable and controlled manner than with run-of-the-river hydropower.

Many developed countries, such as the U.S., Canada, and Norway, witnessed substantial growth in hydropower development in the 1950s and 1960s. Rapid economic growth and industrial development in developed countries in the U.S., Europe, Japan and Australia drove the growth of the hydropower industry until 1975. During this period, heavy industries flourished adjacent to hydropower plants to optimize water resources and minimize power transmission costs.

During 1975–1990, developing countries in Asia and Latin America increased their shares in global hydropower expansion, while the share of additions in developed countries dropped below 50% for the first time. The Itaipu Dam hydropower plant was built in South America in 1984 at 14 GW.

This geographical shift has intensified since 1990, with 85% of new hydro plants being constructed in developing countries. At the same time, as one of the oldest green energy alternatives for fossil fuels, hydropower plays a significant role in energy transition and carbon reduction. In this period, China demonstrated remarkable leadership in the global hydropower market, with a 50% share in global gross capacity expansion, and the Three Gorges Dam hydropower plant in China became the world's largest hydropower plant at 22.5 GW in 2021.

B. **Latest Trends**

At the same time, the deployment of hydropower presented several trends. First, as a mature power generation technology, the further innovation and deployment of hydropower have been limited. In the past two decades, the global installed hydropower capacity rose by more than 69%, from 823 GW in 2003 to 1407 GW in 2022, accounting for an average annual growth rate of 3.5%.

Figure 4.5 shows an overview of this development. It also indicates that since 2013, in which the annual growth rate peaked at 4%, the annual growth rate has continued to decline to 1% in 2019, partly because of the COVID-19 pandemic. However, the growth rebounded to 2% in the following two years and recovered to 3% in 2022.

In the cross-country comparison, the composition of the shares of installed hydropower capacities of the top 10 countries largely remained unchanged in the

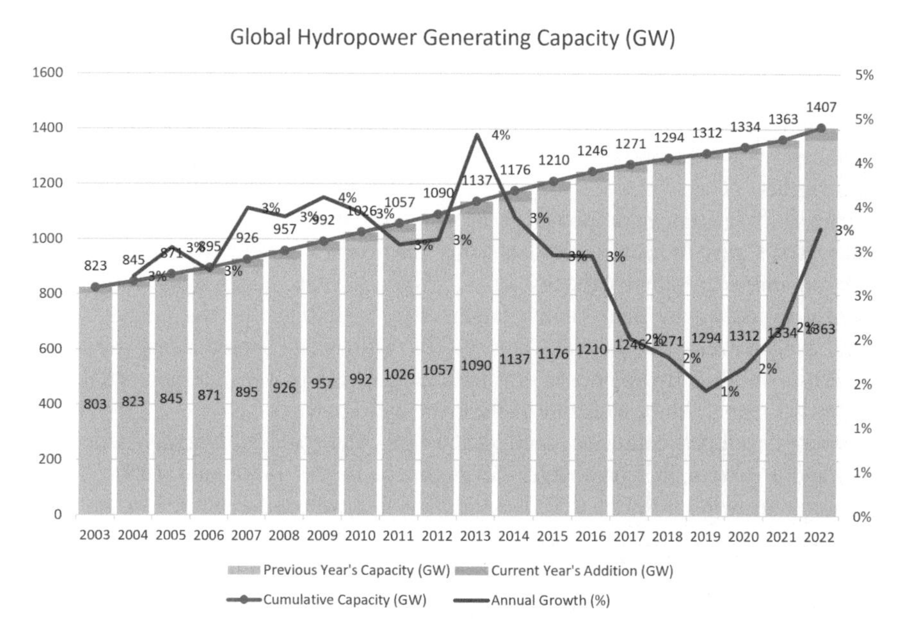

Fig. 4.5 Global hydropower generation capacity growth (2003–2022). *Data Source* IRENA, 2023

last decade from 2013 to 2022, with the exceptions of China, the U.S., and the rest of the world. China increased its share of hydropower capacity in the global hydropower capacity by 4%, from 25% in 2013 to 29% in 2022, while the U.S. reduced its share by 2%, from 9% in 2013 to 7% in 2022, and the rest of the world reduced its share by 1% during the same period (Fig. 4.6).

In addition, hydropower has been the largest renewable energy source since its first invention in the 1870s. In 2022, hydropower represented 42% of all renewable energy. No other renewable energy is expected to overtake its leading role in renewable energy deployment in the short term. With 47 GW of newly installed capacity, the global hydropower capacity rose to 1407 GW and generated 4311 TWh of clean power in 2022, accounting for 15% of the global power supply.

Third, the global deployment of hydropower was extremely unbalanced. Approximately 80% of new hydropower installations in 2022 came from a single country—China. Its hydropower capacity has rapidly expanded in recent years, although its contribution to total power generation has shown a slightly decreasing trend. It overtook Brazil as the world's top hydropower generator in 2001. Since then, it has been leading the global deployment of hydropower technology except for a couple of years, and its leading gap has been expanding since it took off in 2004. As of the end of 2022, China's hydropower reached 414 GW, which accounted for approximately 29% of the world's total hydropower capacity and more than the combined hydropower capacity of the next five countries—Brazil,

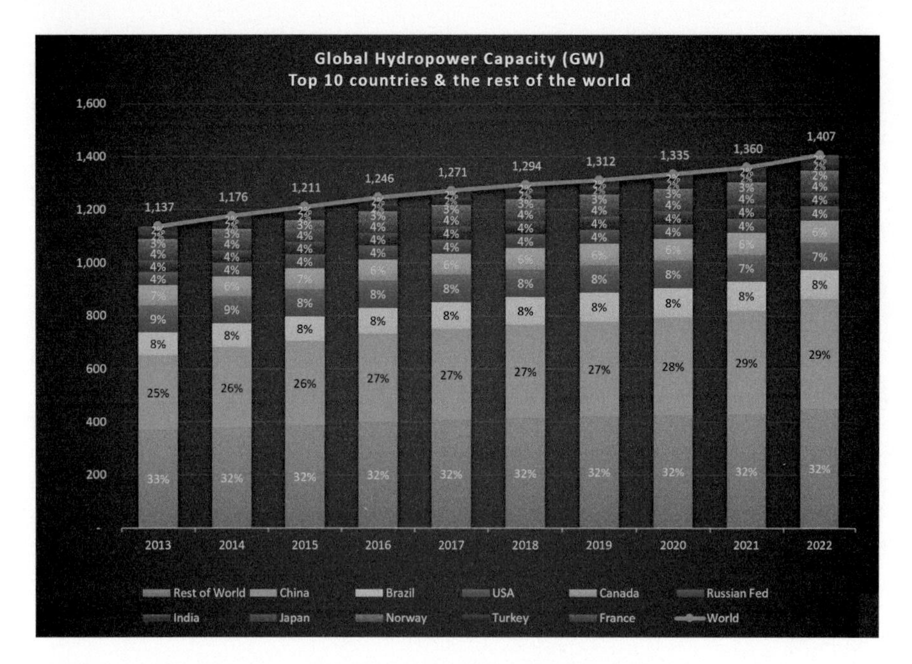

Fig. 4.6 Global hydropower (2013–2022). *Data Source* IRENA, 2023

the U.S., Canada, Russia, and India (Fig. 4.7). Hydropower accounted for 16% of China's power generation and 19% of China's installed power capacity.

Fourth, the share of hydropower in a country's total power generation (hydropower generation) does not necessarily just rely on its development level but also depends highly on its energy demand and natural resource endowment of water resources. For example, developed countries such as Norway had very high hydropower penetration rates of 88.3%.

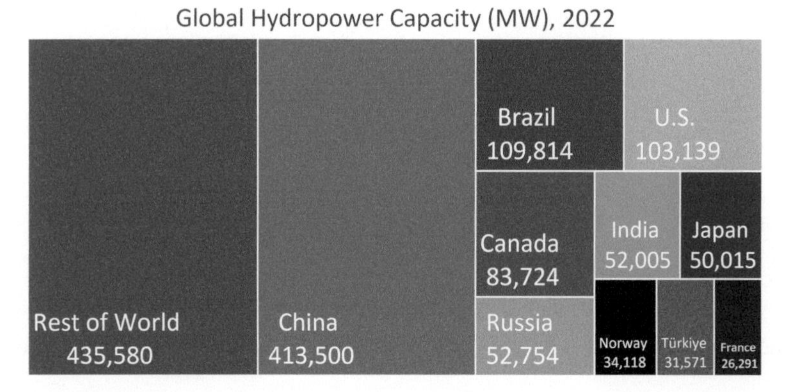

Fig. 4.7 Global hydropower capacity by country (2022). *Data Source* IRENA, 2023

However, many small developing countries, such as Bhutan, Nepal, Albania, Zambia, Paraguay, and Kyrgyzstan, have even higher hydropower penetration rates ranging between 86 and 99%. Globally, 35 countries enjoy hydropower as a major electric power source that cost-effectively delivers more than 50% of power generation. Of these countries, 28 are developing economies with a total population of 800 million. In the developed economies of Norway, Canada, Switzerland and Austria, hydropower provides most of the power supply for decades (Fig. 4.8).

C. **Hydropower Plants in the World**

Approximately 62,500 hydropower plants are generating power worldwide, with a total installed power generation capacity of 1407 GW in 2022.

Hydropower plants vary in size from large power plants, which supply power to consumers and businesses in a large area, to small, mini, micro, and even pico plants, which are operated by local communities or individuals for their own energy needs or for selling power to utilities. There is no unified classification of the sizes of hydropower plants. The following specifications are accepted. The general ones are more defined and used in different countries, and the U.S. Department of Energy uses less detailed specifications for hydropower plants in the U.S. (Fig. 4.9).

Large hydropower plants. As of 2022, 199 hydropower plants in the world have a power generation capacity larger than 1 GW. The majority (168 or 84%) of these large hydropower plants have a power generation capacity between 1 and 3 GW, 20 (10%) of them have a capacity between 3 and 6 GW, and 11 (6%) of them have a capacity larger than 10 GW (Fig. 4.10).

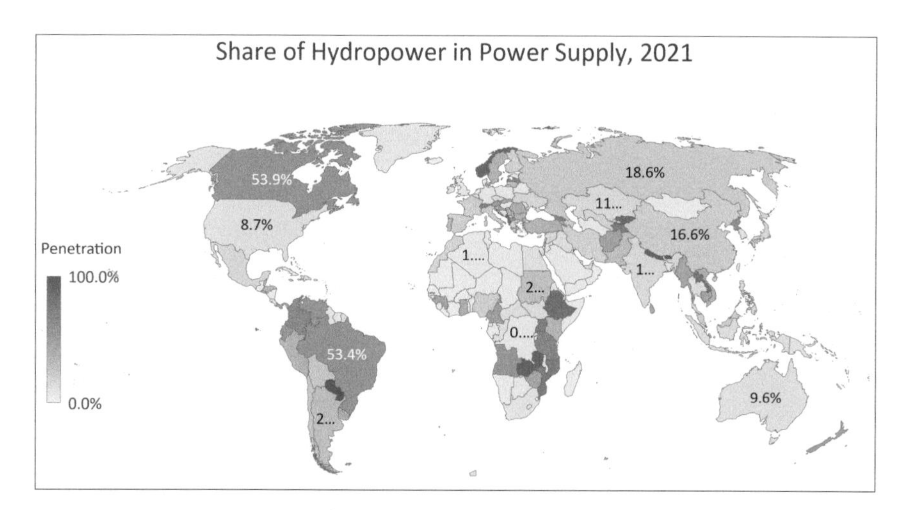

Fig. 4.8 Share of hydropower in total power generation. *Data Sources* IRENA, GlobalEconomy.com

	Large	Small	Mini	Micro	Pico
General	> 50 MW	< 50 MW	<1 MW	< 100 kW	< 10kW
US DOE	> 30 MW	100 kW–10 MW		< 100 kW	

Fig. 4.9 Size of hydropower plants

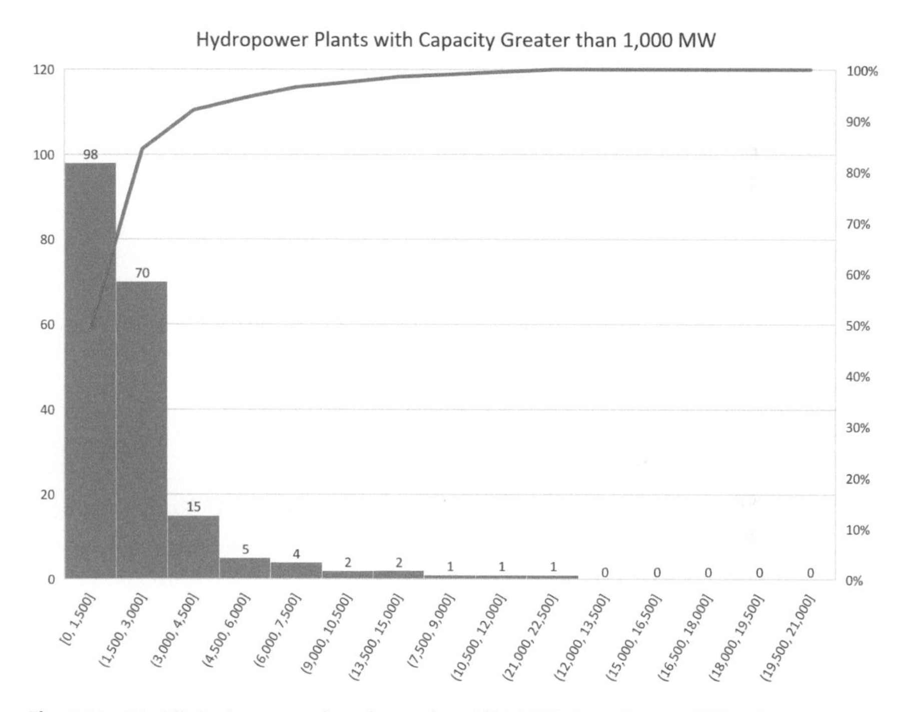

Fig. 4.10 World's hydropower plants larger than 1000 MW. *Data Source* Wikipedia

The world's seven largest hydropower plants have also been its largest power plants since 2022. Four of them are in China. The largest hydropower plant, the Three Gorges Hydropower Plant, is on the Chang Jiang (Yangtze River) in China. It has an installed capacity of 22.5 GW and generated 185.6 TWh in 2022. The second largest hydropower plant, the Baihetan Hydropower Plant, is on the Jinsha River, i.e., the upper course of the Chang Jiang. It has an installed capacity of 16 GW and an annual power generation of more than 60 TWh. The third largest hydropower plant, the Itaipu Hydropower Plant, is in South America on the Paraná River, which forms the border between Brazil and Paraguay. This hydropower plant is owned jointly by the two countries. It has an installed capacity of 14 GW and 103.1 TWh of power generation.

Name	Country	River	Capacity (GW)	Annual Production (TWh)	Area Flooded (km^2)	Reservoir Volume (km^3)	Year of Completion
Three Gorges	China	Yangtze	22.5	111.8	1,084	39.3	2012
Baihetan	China	Jinsha	16	60.24	132	20.62	2022
Itaipu	Paraguay, Brazil	Paraná	14	103	1,350	29	2003
Xiluodu	China	Jinsha	13.86	55.2	108	12.67	2014
Belo Monte	Brazil	Xingu	11.23	39.5	441	1.89	2019
Guri	Venezuela	Caroní	10.24	53.41	4,250	135	1986
Wudongde	China	Jinsha	10.2	38.91	127	7.4	2021

Fig. 4.11 Seven large hydropower plants. *Data Source* Wikipedia, 2023

The fourth largest hydropower plant, the Xiluodu Hydropower Plant, is also on the Jinsha River. It has an installed capacity of 13.9 GW and an annual power generation of 55.2 TWh. The fifth largest hydropower plant, the Belo Monte Hydropower Plant, is on the Xingu River in Brazil. It has an installed capacity of 11.2 GW and an annual power generation of 39.5 TWh. The sixth largest hydropower plant, the Guri Hydropower Plant, is on the Guri River in Venezuela. It has an installed capacity of 10.24 GW and an annual power generation of 53.4 TWh. The seventh largest hydropower plant, the Wudongde Hydropower Plant, is on the Jinsha River in China. It has an installed capacity of 10.2 GW and an annual power generation of 38.9 TWh (Fig. 4.11).

Among the seven largest hydropower plants, Itaipu has the highest *capacity factor*[1] of 84% (rate of power generation by unit of capacity), Guri (60%) and Three Gorges (57%) follow, and all other four have less than 50% capacity factors. In addition, Guri has the highest *unit flooded area*[2] per power generation (79.6 km^2 per TWh), the other two Latin American hydropower plants follow—Itaipu 13.11 km^2 and Belo Monte 11.2 km^2, and the Chinese hydropower plants have lower unit flooded areas—Three Gorges 9.7 km^2 and all other three below 4 km^2. The higher capacity factors of the South American hydropower plants indicate that they might have higher water availability, lower seasonal variations, and more efficient power generation than the Chinese hydropower plants. Higher unit flooded areas of the South American hydropower plants indicate that they might have higher environmental impact than the Chinese ones.

[1] The ratio of actual power generation to the maximum possible power generation if the plant operated at its full capacity for a specific time period (usually a year).

[2] The land area flooded (in km^2—square kilometers) per unit of power generated (TWh—Terawatt-hour).

Challenges

A. Technological Challenges

1. Geographic Variation

The best location for a hydropower station should be along the path of a river. It should be at least at the river canyon or at the place where the river narrows. This enables the collection of water or the diversion of the river, such as mountain gaps that funnel and intensify the energy in waters, which are normally more available in mountainous regions. However, the geographic distribution of water resources is uneven. The global hydropower potential maps show that large parts of the Western U.S. and Southwest China have far more hydropower potential than the rest of the countries. The uneven water resource distribution can be a major challenge to the further expansion of hydropower in those areas. The geographic mismatch of abundant water resources and economic centers causes high curtailment rates of hydropower in remote water-rich regions because a large amount of power generated cannot be consumed locally and requires additional infrastructure to transmit hydropower across the country. Transmitting enormous amounts of variable hydropower to economic centers adds challenges and costs to hydropower penetration. In addition to geographic location, hydropower energy varies significantly depending on the geographic conditions of the site, the weather conditions, and the seasons.

2. Seasonal Variation

The flow in a stream or river can vary significantly from season to season. Therefore, the operation of small run-of-the-river hydropower plants that largely depend on variable water resources can be unstable and unpredictable. At the same time, small hydropower projects require a relatively mild climate to be fully effective—frozen rivers produce no power, while rivers swell with excessive precipitation floods and are destroyed. Hydropower with dams and reservoirs enjoys a more stable water source for power generation as seasonal variations in water flow are balanced. However, dams and reservoirs mean artificial alteration of natural water flow and have significant social and environmental impacts.

3. New Project Challenges

Initiating new hydropower projects necessitates substantial capital expenditure and a commitment to an extended period of meticulous planning. Harnessing water resources for power generation extends beyond a mere short-term venture; it encompasses a wide-ranging investment timeline that might span generations. The construction of a fully equipped hydropower plant, including a dam and reservoir, is not just an engineering feat but an intricate exercise in economic strategy, requiring long-term vision and adept planning over numerous decades.

However, these ambitions often meet substantial social and political obstacles. Social resistance, fueled by concerns over environmental impacts and displacement, can significantly hinder project progress. In the U.S., the permitting process can be complex and time-consuming, often taking several years. For example, the licensing process for a noncontroversial project at an existing dam typically takes approximately 5–6 years, but for more complex projects, it can take up to 10 years or more. In the EU, the length of time required for permitting can also vary widely between member states due to differing national regulations and procedures. It can take anywhere from a few years to more than a decade. For example, in France, the process is usually shorter, taking approximately 2–3 years, whereas in Germany, it can take up to 10 years. In Australia, hydropower projects are typically assessed at the state level, and the process can take between 3 and 5 years, depending on the complexity and scale of the project. In many developing countries, the permitting process can also be lengthy, particularly when it involves large-scale hydropower projects. For example, in India, large projects can take up to 7 years or more to receive all the necessary permissions and clearances. Meanwhile, developing nations often face their own unique challenges, notably the paucity of funds necessary for such large-scale infrastructure projects.

4. Ecosystem Impacts

Hydropower can also cause environmental and social problems through dams, reservoirs, and turbines. Poorly planned hydropower can cause more climate problems than it prevents. Hydropower plants need large dams and reservoirs to provide a steady flow of water. Constructing reservoirs will flood plants and other organic matter. This submerged vegetation decays over time, releasing greenhouse gases such as methane and CO_2.

Dams and reservoirs drastically affect the landscape and rivers. They can reduce river flows, raise the water temperature, degrade water quality, and cause build-ups of sediment. This has negative impacts on fish, birds, and other wildlife. Large dams and reservoirs have major negative impacts on river ecosystems because they prevent some animals from traveling upstream, cool and deoxygenate water discharged downstream, and reduce downstream nutrients because of settling particulates. They cause emissions of greenhouse gases, especially methane, from decomposing accumulated vegetation at the bottom of the reservoir. In addition, they can cause habitat loss for some aquatic species.

Hydropower facilities, such as turbines, spillways, and other infrastructure, can have a range of impacts on fish populations, including mortality, injury, displacement, and altered behavior. Fish can be killed or injured by collisions with turbine blades, rocks, or other structures at hydropower facilities.

The construction of dams and other hydropower infrastructure can alter the natural flow patterns of rivers and streams, which can affect the distribution and migration of fish populations. This can lead to changes in the abundance and diversity of fish species in the affected areas.

5. Social Impact

Large hydropower plants need large dams with large reservoirs. As a result, people who live around reservoirs are relocated or displaced. This is especially true for building a massive hydropower dam in a heavily populated area. Another potential impact is that cultural or religious sites may be submerged by reservoirs. Finally, dam failures can also have calamitous impacts, such as displacement and loss of life and property. A World Bank study estimated in 2000 that building dams and reservoirs directly displaced 40–80 million people. Another study in 2010 found that 472 million people downstream from large dams suffer from reduced food security, regular flooding, or impacts on their livelihood.

Therefore, while hydropower is undeniably a low-carbon power source, every country with abundant untapped water resources needs to carefully balance the benefits of hydropower with the potential environmental and social costs of dam projects. There are significant needs and potential for hydropower development. However, careful and proactive investigation and planning are needed to minimize the environmental and social costs of hydropower projects.

6. Declining Hydropower Penetration

Despite the urgent need for renewable energy to meet the IEA's 'Net Zero by 2050' scenario and the apparent achievements of rising global hydropower generation, the share of global hydropower in total power generation was only 15.2% in 2022. In addition, this has been a continuation of a long-term downward trend from 20% in 1986 (Fig. 4.12).

The declining trend of the share of hydropower is caused by the diversification of energy sources in the last century, the faster declining growth of hydropower in developed countries than in developing countries, and the lower hydropower shares in total power generation in most developed countries than in most developing countries.

It is worth noting that while developed countries, such as the U.S. and Germany, had high shares of hydropower (i.e., a high share of hydropower in the total power generation) in the past, their hydropower shares have significantly decreased and their hydropower generation stagnated over time, although their hydropower generation has significantly increased. For example, U.S. hydropower generation rose by 1.6-fold from approximately 101 TWh in 1950 to more than 262 TWh in 2022. However, its hydropower share dropped by approximately 80% from more than 30% in 1950 to 6.17% in 2022. This means that the importance of hydropower for the U.S. economy has decreased (see Fig. 4.13).

However, the Net Zero scenarios require global hydropower to double its current installed capacity by 2030 and to help renewable energy sources, together with solar power and wind power, triple their power generation and provide 90% of the total power supply by 2050. Considering that global hydropower has increased by only 25% in the last decade, meeting the Net Zero goal is an extremely challenging task for global hydropower.

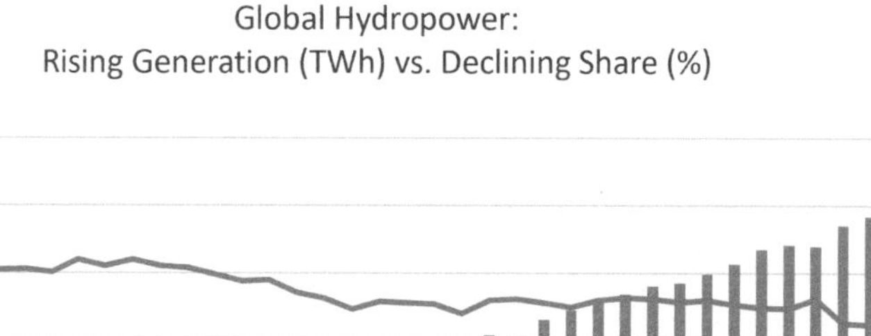

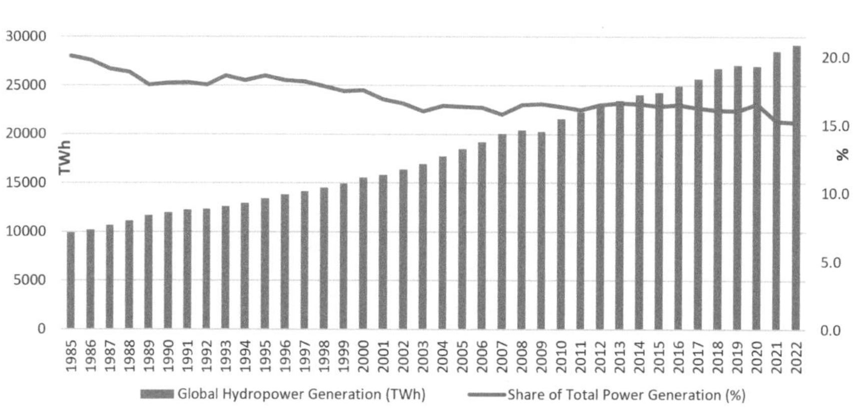

Fig. 4.12 Changes in the share of hydropower in global power generation. *Data Sources* World Bank, Our World in Data, etc.

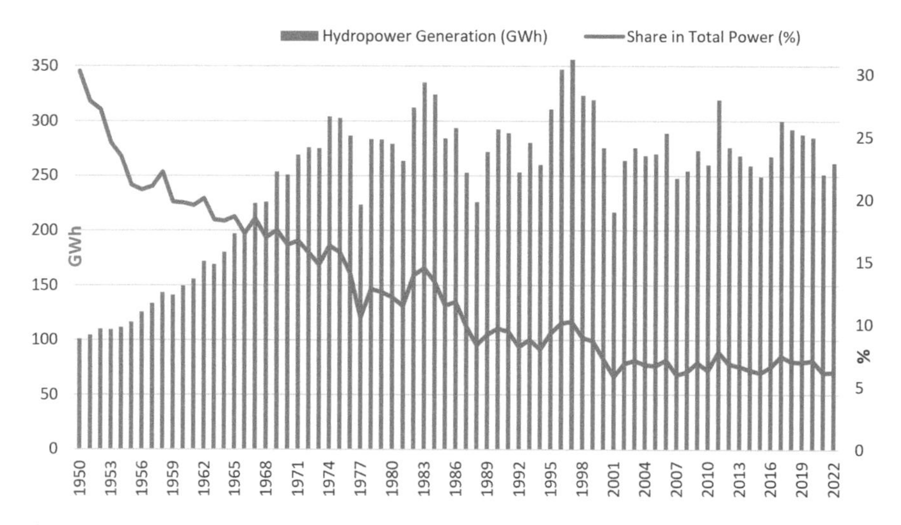

Fig. 4.13 U.S. hydropower. *Data Source* U.S. Department of Energy, 2023

7. Curtailment

Power curtailment has been a major issue for large new hydropower plants in Southwest China. The curtailment rate in hydropower plants on the Dadu River accounted for 53% of the entire province of Sichuan, causing it to be at the top of the list of provinces affected by hydropower curtailment. From 2015 to 2019, the hydropower curtailment of the power plants along the river exceeded 40 GWh. The main causes of hydropower curtailment include the rapid increase in newly installed hydropower generation capacities and the lack of a power transmission grid, especially the ultrahigh voltage direct current transmission grid that can be used to move power from new hydropower plants to power-consuming East China coastal regions.

Seasonal overproduction of hydropower can also cause curtailment of power generated from other sources, and renewable energy, especially utility-scale solar PV or wind power, could be the victim of curtailment. The profusion of hydropower occurs when enormous amounts of snowmelt and rain pour into reservoirs. In California, for example, the potential curtailment amount is estimated to be between 6 and 8 GW, an equivalent output of 6–8 nuclear power plants. Rooftop solar PV power that is not fed into the power grid will not be affected. The power grid must accept power from hydropower plants if water is above their safety threshold ("spill levels") and must be discharged. Hydropower can also be curtailed when water is beneath dam spill levels.

8. Aging Dams

Despite the low shares of hydropower in the total power generation of some developed economies, such as the U.S. (6.1%), Germany (3%), and the UK (1.8), there are many old dams and reservoirs. The U.S., for example, has over 91,986 dams that serve many purposes. Of this substantial number of dams, only approximately 3% generate hydropower, many of these hydropower plants are owned by private utilities, and their operations and maintenance follow rigorous rules.

However, many dams in developed countries are incredibly old. In the U.S., the average age of dams is 61 years, and the share of dams older than 50 years will reach 70% by 2030. These dams are classified by hazard potential in the U.S., and the number of high-hazard-potential dams has more than doubled in the last two decades with increasing development close to rural dams and reservoirs. There is an Emergency Action Plan to protect these dams, and more than 81% of the high-hazard-potential dams had a protection plan on file. However, because of a lack of funding, many of the protection plans could not be implemented, and the number of deficient high-hazard-potential dams surpassed 2300.

9. Inadequate Transmission Grid

Transporting utility-scale hydropower from remote onshore and offshore hydropower farms to urban load centers where most of an economy's energy is consumed requires an adequate transmission grid. Increasingly higher penetration of intermittent hydro resources and power output poses unique problems in transmission planning and efficient utilization of transmission infrastructure, resulting in higher transmission losses and costs, increased congestion, and even power curtailment with a lack of adequate transmission capacity. Due to potential transmission constraints, hydropower project developers will need to evaluate the economic tradeoff of sitting where the resource is best versus sitting closer to loads where transmission constraints are less likely.

10. Climate Change

Hydropower relies on the world's river basins and hotspots for its future development. However, in the coming decades, climate change is expected to have significant impacts on the deployment of hydropower. Climate change-driven deforestation, droughts, and reductions in glaciers, precipitation, and river discharge will significantly impact the hydropower capacity in those river basins.

First, changes in precipitation patterns and temperatures may alter the availability of water resources, affecting the amount of energy that can be generated by hydropower facilities. Droughts, for example, can reduce water levels in rivers and reservoirs, decreasing the capacity of hydropower facilities to generate electricity.

Second, extreme weather events such as floods and storms may damage hydropower infrastructure and affect the reliability and safety of hydropower facilities. For example, heavy rainfall and flooding can cause erosion and sedimentation, potentially impacting the efficiency of turbines and other equipment.

Solutions

Hydropower plays a key role in the energy transition, as it provides cheap and reliable renewable energy and energy storage and counterbalances other intermittent renewables. Maximizing the benefits of hydropower and minimizing its challenges has become increasingly important because all renewable energy sources need to grow to stop intensified climate change, and growing mature and inexpensive hydropower technology is more cost-effective than other renewable energy sources.

A. **Revitalizing and Repurposing Existing Dams**

One of the major effective solutions is to revitalize and repurpose existing dams for hydropower or PHS. This is especially true for developed countries such as the U.S. and Germany, where a substantial number of dams were built but are currently unused. Dams are useful for pumping ton of renewable energy to the grid. According to the American Society of Civil Engineers (ASCE) 2021 Infrastructure Report Card, there are 91,000 dams in the U.S., of which only 3% or less than 2300 generate power as of 2021. The other dams are used for recreation, stock/farm ponds, flood control, water supply, and irrigation. In addition, approximately 17,000 dams (19%) are considered high-hazard potential dams because their failure could result in the loss of life or severe damage to properties and infrastructure downstream.

Currently, a typical approach to old dams is dam removal. However, the removal of dams also has short-term and long-term negative impacts. Short-term impacts include increasing water turbidity and sediment buildup downstream from releasing substantial amounts of sediment from the reservoir and deteriorating water quality from sudden releases of water and changes in temperature. Long-term impacts include unacceptable social, environmental, and economic outcomes, such as the loss of flood protection for populated areas, the loss of wetlands created by the dam, or the loss of energy from a hydropower project.

Modification of old dams can improve stream water quality, reintroduce aquatic species up- and downstream from the site, and increase recreation while still maintaining some control of flows. Dam modification is primarily practiced with low-head dams, as they pose the most significant safety risk. Transforming the structure's design reduces the risk of drowning by altering the water's vertical drop and cyclical undercurrents. Nevertheless, it can be applied to other dam varieties, depending on size, location, and overall condition.

The great benefits of revitalizing and repurposing the existing dam infrastructure include avoiding the need for building new dams, saving major construction costs, and significantly reducing the investment in adding installed hydropower capacities. In addition, vitalizing and repurposing projects can also improve the safety and health of dams, increasing dams' social benefits without adding new environmental costs to river ecosystems and new social costs to populated areas. To achieve this goal, its potential needs to be analyzed and assessed using a spatially comprehensive global inventory of geolocated dams used for purposes other than hydropower, and the results are enhanced using modeled estimates of small, unmapped dams. In addition, increases in hydropower potential from efficiency upgrades at existing hydropower plants need to be examined. This kind of project can offer noninvasive additions to hydropower capacity to displace fossil fuel-fired power plants in populated areas. Some studies estimate that this kind of project would potentially add up to 9% of the current global hydropower capacity or pumped hydropower storage capacity, with potentially reduced costs of construction and transmission and reduced potential impacts on biodiversity and river ecosystems by avoiding building new hydropower dams.

B. **Small (Mini, Micro, and Pico) Hydropower**

To increase sustainable hydropower generation and maximize future hydropower potential, it is extremely important to facilitate small hydropower and pumped storage projects. A study published in 2018 revealed that 82,891 SHPs were operating or under construction, which means that the ratio of small to large hydropower plants is 11:1 and that this number is estimated to double if all potential generation capacities are to be developed. These projects can be especially sustainable and beneficial if they are planned and carried out in conjunction with existing dam protection programs and plans. With additional investment and government support, old dams can be improved by renovation and repurposing for power generation. As a result, they can be rejuvenated and thus environmentally safe and more economical because the combined projects can avoid significant amounts of funding for building new dams, the revenue from power generation can recover the investment in projects, and the completed projects will provide more clean power and more revenue for the local communities.

More government support is necessary for the development of innovative hydropower technologies on sites that were previously deemed impractical. For example, testing and using hydrokinetic units on small ponds, creeks, and waterfalls can generate power in a free-flow environment with no need for a dam or canal drop. These water resources are small in terms of power generation capacity, but they can generate enough power for an isolated family, small community, or rural industry in remote areas.

C. **Smart Hydropower Technology**

Since the industrial revolution, hydropower has evolved from passively harnessing water to generate hydropower into a sophisticated and complex science since the early projects of using water energy on rivers. Currently, weather satellites and related weather forecasting systems are used to predict precipitation and stream flows, and smart digital algorithms are utilized to calculate the discharge of water from reservoirs for hydropower generation in combination with a range of other needs, including irrigation needs, domestic and recreational needs, flood control, navigation, and fish and wildlife conservation.

Smart hydropower or digitization of hydropower is an important approach to the challenges facing hydropower technology. Smart hydropower covers the digitization of hydropower turbines and hydropower plants, including technologies assessing and managing hydropower turbines, reservoirs, and water resources using high-fidelity computer modeling research, the physics surrounding hydro turbine noise generation, integrated hydropower plant control, system design, system analysis, and reliability. Using digitization and AI tools to measure, record, and analyze water data, smart hydropower technologies will allow hydropower plant operators to achieve enhanced power generation and more efficient material use; reduce operation, maintenance, and service costs; extend hydro turbine and hydropower plant life; and optimize hydropower grid control and reliability.

D. **Fish-Safe Pathways and Turbines**

To protect fish passage from the impacts of hydropower facilities, governments around the world, especially in developed countries, have established fish-friendly regulations and guidelines, including requirements for the design and use of fish-friendly turbines such as low head turbines and modified Kaplan turbines. In other parts of the world, such as Asia and South America, fish-friendly turbine technology is less widespread, but there is growing interest in developing and implementing these designs to reduce the impact of hydropower on fish populations.

Designing and utilizing fish-friendly turbines and pathways can help reduce the impact of hydropower facilities on fish populations. Currently, there is no universally accepted benchmark for fish survival rates of fish-safe turbines. However, some studies have suggested that a minimum survival rate of 95% is necessary for a turbine to be considered safe for fish.

Fish-safe hydro turbines are designed to be as efficient as existing high-efficiency turbines. In fact, some fish-friendly turbine designs have been shown to improve overall turbine efficiency while also reducing the impact on fish populations. For example, low head turbines, which are designed to operate at low head sites (i.e., sites with a small drop in elevation), are typically more fish-friendly than traditional high-head turbines. These turbines have been shown to be as efficient as traditional turbines while also reducing the risk of injury or mortality to fish. Similarly, Kaplan turbines with modifications to reduce blade strike and water pressure can also improve the safety of fish passage.

Fish-friendly modifications have been made to turbines to improve fish passage and reduce mortality. These modifications can include the reduction of the number of fish-friendly nozzles, screens, or deflectors.

Fish-friendly *nozzles* are designed to reduce the risk of blade strike and pressure changes for fish passing through the turbine. Modifying the shape and size of the nozzle can reduce water turbulence and provide a smoother flow of water through the turbine. The use of fish-friendly nozzles has shown an increase in fish survival rates through the turbine.

Installing *screens* in front of the turbine can prevent fish from entering the hydro turbine. The screens can be designed with a variety of mesh sizes and shapes to allow water to pass through while preventing fish from being pulled into the turbine. The use of screens has shown reduced fish mortality.

Installing *deflectors* inside the turbine can create a smoother flow of water through the turbine, redirect water flow, and guide fish away from the blades. This design can reduce the risk of blade strike and pressure changes for fish passing through the turbine. The use of deflectors has been shown to increase fish survival rates through the turbine.

Installing *fish ladders*, rock weirs, or bypass channels are designed to provide additional options for fish passage and to reduce mortality.

A *fish ladder* is a structure that helps fish past obstacles such as dams or other barriers by swimming and leaping up a series of relatively low steps (i.e., "ladder") into the waters on the other side upstream. Fish ladders typically consist of a series of pools or steps, with water flowing over them to create a current that helps fish swim up and around barriers. The pools or steps are designed to provide resting areas for fish, which need to conserve energy as they swim upstream. The velocity of water running down the steps should be not greater than 2.3 m/s so that fish will not be washed back downstream or exhausted for the upstream journey. Fish ladders are successfully used in many hydropower facilities around the world to help fish migrate upstream for spawning or other purposes. However, the effectiveness of fish ladders depends on a variety of factors, including the design of the ladder, the species of fish, and the flow of water. Some fish species are less successful in using fish ladders than others.

A *rock weir* is a structure made of rocks or boulders in a river or stream below a dam, designed to create a shallow area in the river to help fish migrate upstream and improve fish habitat by creating areas with slower-moving water. Rock weirs can improve fish habitat and facilitate fish passage, particularly for smaller fish species. The effectiveness of rock weirs can depend on several factors, including the design and placement of the structure, the size and species of fish, and the flow of the water. This structure is particularly effective for fish species that prefer shallow water, such as salmonids or other small fish species. Building a rock weir below the dam can greatly reduce fish mortality. In many cases, the costs of new rock weirs can be much lower than those of new fish ladders.

Summary

Hydropower is a mature renewable energy generation technology that is used to generate power by mechanically converting aquatic energy into power. This technology differs from power generation technologies using fossil fuels. It uses free river water instead of fossil fuels to generate power and is as safe as solar power, safer than wind power, and much safer than fossil fuels. Although it is not as clean as solar and wind power, it has immense potential to offset most CO_2 emissions from fossil fuel-fired power generation and meet the global Net Zero target. For more than 20 years, government support in many countries has helped hydropower manufacturers produce ever larger and increasingly capable hydro turbines through R&D and incentivized hydropower plant operators to deploy hydropower technologies onshore and offshore. The continued innovation and economies of scale in hydropower led to hydropower's competitive edge over the cheapest coal-fired power. However, hydropower generation still faces many challenges, especially in offshore expansion, and solutions are urgently needed for its full penetration to meet the global Net Zero goal by 2050.

In the current hydropower market, large hydropower plants offer higher efficiency but also present larger social, environmental, and ecological impacts. Small hydropower plants have lower social, environmental, and ecological impacts but are less efficient and cost more per kW of power generated. To address the challenges of hydropower development, developed countries with much more developed water reservoirs and dams need to focus on renovating, modifying, and repurposing existing dams, and developing countries with more untapped water resources but fewer financial assessments need to consider hydropower solutions that can reduce social, environmental and ecological impacts when they endeavor to substantially develop hydropower reservoirs and dams.

There is no single panacea that can solve the multiple challenges of hydropower. To reduce the high hydropower investment and maintenance cost, more proactive measures need to be taken, for example, working out stringent policies on assessing the environmental and ecological impacts of hydropower generation;

educating consumers on price comparison with fossil fuel-fired power in terms of avoided fuel cost and external cost; providing government funding for new hydro turbine R&D and continued deployment of digitization and AI technology in the smart hydro turbine, smart hydropower plant, and smart hydropower grid; expanding hydropower generation, streamlining review process to reduce red tape and delays; and improving efficiency in permitting and licensing, providing hydropower investors with financial incentives or commercial loan programs.

Activities

Further Readings

1. The changing role of hydropower: Challenges and opportunities. https://www.irena.org/Publications/2023/Feb/The-changing-role-of-hydropower-Challenges-and-opportunities
2. Hydropower in China – Development and Challenges. https://blog-isige.minesparis.psl.eu/2020/05/14/hydropower-in-china-development-and-challenges/
3. Assessment of Challenges and Opportunities in the Hydropower Sector in Nepal. https://www.linkedin.com/pulse/assessment-challenges-opportunities-hydropower-sector-tejaswi-sharma/
4. Hydropower Opportunities and Challenges. https://www.doi.gov/ocl/hydropower-opportunities-and-challenges
5. Modernizing Energy Infrastructure: Challenges and Opportunities to Expanding Hydropower Generation. https://www.congress.gov/event/115th-congress/house-event/105702/text
6. Hydroelectricity: Major Challenges and Issues. https://www.fuergy.com/blog/hydroelectricity-major-challenges-and-issues
7. Hydropower Supply Chain: Potential Solutions to Complicated Challenges. https://www.hydro.org/powerhouse/article/understanding-the-hydropower-supply-chain-opportunities-challenges-and-potential-solutions/
8. Giving old dams new life could spark an energy boom. https://www.washingtonpost.com/climate-solutions/2022/05/06/dams-nonpowered-hydropower-energy/
9. Considering the Alden Turbine for a Plant Rehab. https://www.hydroreview.com/world-regions/considering-the-alden-turbine-for-a-plant-rehab
10. Emerging technologies and challenges in hydropower. https://www.prescouter.com/2019/10/hydropower-emerging-technologies-and-challenges/
11. Hydropower Modeling Challenges. https://www.nrel.gov/docs/fy17osti/68231.pdf
12. Challenges and Opportunities for New Pumped Storage Development. https://www.hydro.org/wp-content/uploads/2017/08/NHA_PumpedStorage_071212b1.pdf

Closed Questions

1. What is hydropower and how does it generate electricity?
   (a) Hydropower is the energy harnessed from sunlight
   (b) Hydropower is the energy harnessed from wind turbines
   (c) Hydropower is the energy harnessed from the movement of water, which drives turbines to produce electrical energy
   (d) Hydropower is the energy harnessed from nuclear reactions
   (e) Hydropower is the energy harnessed from burning fossil fuels.
2. What are the distinct types of hydropower systems, and how do they differ from each other?
   (a) Run-of-the-river and tidal power; they differ in the source of water used
   (b) Solar power and geothermal power differ in their renewable energy sources
   (c) Wind power and biomass power; they differ in the type of turbines used
   (d) Coal power and natural gas power differ in the type of fuel burned
   (e) Hydrokinetic power and pumped storage power differ in their storage capabilities.
3. How does hydropower contribute to renewable energy goals?
   (a) Hydropower is not considered a renewable energy source
   (b) Hydropower is only a small contributor to renewable energy goals
   (c) Hydropower provides a reliable and constant source of renewable energy
   (d) Hydropower contributes to renewable energy goals by reducing greenhouse gas emissions
   (e) Hydropower contributes to renewable energy goals by promoting energy conservation.
4. What are the environmental benefits of hydropower?
   (a) Hydropower has no environmental benefits
   (b) Hydropower helps reduce air pollution by emitting fewer greenhouse gases
   (c) Hydropower contributes to water pollution and ecosystem degradation
   (d) Hydropower increases reliance on nonrenewable energy sources
   (e) Hydropower has no impact on the environment.
5. What are the economic advantages of hydropower for the energy market?
   (a) Hydropower is expensive and increases energy costs for consumers
   (b) Hydropower creates job opportunities and stimulates local economies
   (c) Hydropower is not financially viable and requires heavy government subsidies
   (d) Hydropower has no economic advantages compared to other energy sources
   (e) Hydropower increases energy market volatility and instability.
6. What are the main challenges associated with hydropower development?
   (a) Limited availability of water resources
   (b) High construction and maintenance costs
   (c) Environmental concerns and impacts on ecosystems

   (d) Limited public support and opposition
   (e) All the above.
7. How does hydropower impact aquatic ecosystems and fish populations?
   (a) Hydropower has no impact on aquatic ecosystems
   (b) Hydropower improves aquatic ecosystems by providing stable water flows
   (c) Hydropower can disrupt fish migration and habitat
   (d) Hydropower increases fish populations and biodiversity
   (e) None of the above.
8. What is not one of the strategies for minimizing the environmental impact of hydropower projects?
   (a) Implementing fish ladders and bypass channels
   (b) Enhancing environmental flow releases
   (c) Conducting environmental impact assessments
   (d) Implementing sediment management techniques
   (e) Promoting fishing activities.
9. How does hydropower integrate with other renewable energy sources in the energy grid?
   (a) Hydropower cannot integrate with other renewable sources
   (b) Hydropower is complementary to other renewable sources, providing consistent baseload power
   (c) Hydropower competes with other renewable sources for grid integration
   (d) Hydropower is the only renewable source that can be integrated into the energy grid
   (e) None of the above.
10. What is the role of pumped hydropower storage in energy storage and grid stability?
   (a) Pumped storage hydropower is not used for energy storage
   (b) Pumped storage hydropower provides a reliable source of continuous power
   (c) Pumped storage hydropower helps balance fluctuations in energy supply and demand
   (d) Pumped storage hydropower is only used for emergency backup power
   (e) None of the above.
11. How does climate change affect the potential for hydropower development?
   (a) Climate change has no impact on hydropower development
   (b) Climate change decreases the potential for hydropower due to reduced water availability
   (c) Climate change increases the potential for hydropower due to increased water availability
   (d) Climate change affects hydropower development by altering weather patterns
   (e) None of the above.
12. What are the social and cultural implications of large-scale hydropower projects?
   (a) Large-scale hydropower projects have no social or cultural implications

   (b) Large-scale hydropower projects often lead to the displacement of local communities
   (c) Large-scale hydropower projects have positive impacts on local cultures and traditions
   (d) Large-scale hydropower projects do not require community involvement or consultation
   (e) None of the above.
13. Are there any emerging technologies or innovations in the field of hydropower?
   (a) There are no emerging technologies or innovations in hydropower
   (b) Emerging technologies in hydropower include underwater turbines and tidal energy
   (c) Innovations in hydropower focus solely on increasing power generation capacity
   (d) Emerging technologies in hydropower aim to reduce environmental impacts
   (e) None of the above.
14. What is the current market outlook for hydropower globally?
   (a) The global hydropower market is declining
   (b) The global hydropower market is stagnant with no growth potential
   (c) The global hydropower market is experiencing steady growth
   (d) The global hydropower market is highly volatile and unpredictable
   (e) None of the above.
15. How does hydropower compare to other renewable energy sources in terms of cost and scalability?
   (a) Hydropower is the most expensive and least scalable renewable energy source
   (b) Hydropower has lower costs and higher scalability than other renewable sources
   (c) Hydropower and other renewable sources have comparable costs and scalability
   (d) Hydropower is less scalable but more cost-effective than other renewable sources
   (e) None of the above.
16. What are the key policy and regulatory considerations for promoting hydropower development?
   (a) No specific policies or regulations are necessary for hydropower development
   (b) Streamlining environmental regulations and permits is the only consideration
   (c) Ensuring fair pricing and subsidies for hydropower is a key policy consideration
   (d) Balancing environmental protection with streamlined permitting processes
   (e) None of the above.

17. What are not the main barriers to hydropower expansion?
    (a) Limited water resources and geographical constraints
    (b) High upfront costs and long project development timelines
    (c) Environmental concerns and opposition from local communities
    (d) Inadequate infrastructure and transmission capacity
    (e) Incorporated sustainable tourism programs.
18. How do community engagement and stakeholder involvement play a role in hydropower projects?
    (a) Community engagement is unnecessary for hydropower projects
    (b) Stakeholder involvement helps address local concerns and ensures project success
    (c) Community engagement only focuses on economic benefits for the local community
    (d) Stakeholder involvement can be ignored in favor of project expediency
    (e) None of the above.
19. What are some examples of successful small-scale hydropower installations around the world?
    (a) Small-scale hydropower installations have no successful examples
    (b) The success of small-scale hydropower depends on regional and local conditions
    (c) Examples include the MHP program in Nepal and micro hydro projects in Indonesia
    (d) Successful small-scale hydropower installations are limited to developed countries
    (e) None of the above.
20. How can hydropower contribute to achieving sustainable development goals and reducing carbon emissions?
    (a) Hydropower has no role in achieving sustainable development goals
    (b) Hydropower contributes to sustainable development by providing clean and renewable energy
    (c) Hydropower increases carbon emissions and is not environmentally friendly
    (d) Hydropower is not a viable option for reducing carbon emissions
    (e) None of the above.

Open Questions

1. How does hydropower differ from power generation using fossil fuels?
2. What is PHS? What is the main difference between PHS and regular hydropower?
3. What is the range of efficiencies or conversion rates of the PHS?

4. Why is hydropower paramount for combating climate change and global warming?
5. What are the main differences between developed countries and developing countries in their development of hydropower?
6. What types of hydro turbines are there?
7. What are the advantages and disadvantages of the distinct types of hydro turbines?
8. What are the requirements for hydropower?
9. What challenges do the requirements for hydropower pose?
10. Why does hydropower installation have high upfront costs?
11. What solutions are needed to help reduce hydropower costs?
12. What potential challenges do large utility-scale hydropower plants pose to other grid-connected renewable energy generation?
13. What solutions are available to address the challenges of large utility-scale hydropower plants?
14. When is hydropower curtailed and when is it not curtailed?
15. What are the causes of hydropower curtailment?
16. What are the possible choices to solve or avoid the problem of hydropower curtailment?
17. What are the benefits of repurposing existing non hydropower dams and reservoirs for hydropower?
18. What are the benefits of dam modification over dam removal?
19. What are the crucial factors of repurposing old dams for hydropower generation?
20. How can existing old reservoirs contribute to expanding hydropower in developed countries?
21. What are the differences between open-loop and closed-loop PHS facilities? Which one is more eco-friendly? Why?
22. What are the challenges facing PHS?
23. What are the viable solutions to these challenges?

Bioenergy

5

Introduction

Science and Technology

1. Bioenergy

Bioenergy is an important renewable energy source harnessed from biomass, i.e., plants or other organic materials that result directly or indirectly from photosynthesis. While biomass can be divided into two groups—traditional biomass and modern bioenergy—the energy transition focuses on the latter because bioenergy is produced using various technologies, which make it much more efficient and friendly to the environment and human health than traditional biomass in production and consumption. The R&D of bioenergy technologies recognizes three developmental groups or generations of bioenergy. Each of the three generations uses its own feedstock and has its own benefits.

First-generation bioenergy encompasses biofuels derived from existing row crops that serve as a source of human food and animal feed. This includes ethanol produced from corn, sugarcane, or wheat, as well as biodiesel derived from vegetable oils such as soybean oil, rapeseed oil, or palm oil. Examples include ethanol derived from corn starch and sugarcane sucrose, as well as biodiesel derived from soybean. The technologies employed for first-generation bioenergy are based on simple biochemical processes, namely, fermentation and distillation. These processes effectively convert starch and sucrose into ethanol and vegetable oil into biodiesel. One advantage of first-generation technologies is their maturity, as they have been extensively developed and are readily adopted. Moreover, their commercialization in the biofuels sector is less costly compared to other advanced technologies, as there is no significant need for extensive research and development. Ethanol can be used in its pure form, known as E100 (100% ethanol), but it is often blended with gasoline at different ratios, such as E10 (10% ethanol), E15

P. Yang, *Renewable Energy*, https://doi.org/10.1007/978-3-031-49125-2_5

(15% ethanol) or E85 (85% ethanol). Among these blends, E10 is commonly used in the U.S., China, Germany, and Australia.

Second-generation bioenergy primarily consists of waste-based biofuels derived from various sources, including kitchen food waste, agricultural waste (such as crop residues, livestock, and poultry manure), forest biomass residues (such as woody biomass, logging residues, shavings, sawdust, and woodchips), and municipal solid waste (such as household garbage). One example of second-generation biofuel is biodiesel, which can be produced from waste vegetable oils, animal fats, or recycled restaurant grease. Biodiesel has been extensively tested and proven to be a renewable and clean-burning fuel for diesel vehicles. Compared to traditional diesel and ethanol, biodiesel offers several advantages. It requires less input cost, provides greater benefits, and releases fewer toxic pollutants and greenhouse gases. Biodiesel can be used in its pure form, known as B100 (100% biodiesel), but it is often blended with diesel at different ratios, such as B2 (2% biodiesel), B5 (5% biodiesel), and B20 (20% biodiesel). Among these blends, B20 is commonly used in the U.S., Canada, Australia, Germany, Sweden, Finland, Norway, and South Africa. The utilization of second-generation bioenergy, particularly waste-based biofuels, presents a significant opportunity to reduce waste, utilize available resources more efficiently, and contribute to a cleaner and more sustainable energy system. However, challenges remain, such as the need for advanced technologies to efficiently convert various waste materials into biofuels and the establishment of proper waste collection and management infrastructure. Ongoing research and development efforts are focused on optimizing conversion processes, improving feedstock availability, and addressing potential environmental and economic concerns to ensure the future expansion of second-generation bioenergy.

Third-generation bioenergy represents a promising advancement in renewable energy. It utilizes nonfood biomass sources, including algae, to produce efficient biofuels through advanced conversion processes. Algae fuel, derived from algae biomass, is a prominent example. It offers resource efficiency, reduced emissions, and potential for large-scale cultivation. Ongoing research focuses on optimizing processes and addressing challenges for its widespread adoption. Third-generation bioenergy, exemplified by algae fuel, holds enormous potential for a cleaner and sustainable energy future.

2. Bioenergy Processing Technologies

From a methodological perspective, six generic biomass processing technologies can be identified.

(a) Fermentation: This process involves the conversion of starch and sugars into alcohols, such as ethanol. Fermentation has a long and rich history, dating back to 10,000 BC.
(b) Anaerobic digestion: This method is used for the decomposition of food waste, including fats, oils, greases, animal manure, industrial organic residuals, and sewage sludge (biosolids), to produce methane-rich gas, which can be used as

a source of biofuel. Anaerobic digestion has been practiced for approximately 2000 years.

(c) Transesterification: Currently, transesterification is being investigated in research and development. It involves the conversion of woody, fatty, and oily biomass into bioenergy.

(d) Direct combustion: This technology is widely employed in developed countries and increasingly adopted in developing countries. It entails the burning of solid waste to generate power and heat.

(e) Gasification: Gasification involves the conversion of solid waste into carbon monoxide and hydrogen-rich syngas. This technology is currently being commercialized.

(f) Pyrolysis: Pyrolysis is a process in which solid waste is subjected to elevated temperatures in the absence of oxygen. This results in the production of biochar, gas, and oils. Pyrolysis is currently being experimented with and demonstrated.

Recognizing these six biomass processing technologies can help researchers and practitioners better understand and explore the various methods available for utilizing biomass as a renewable and sustainable energy source.

While the first three processing technologies are based on biochemical conversion, the last three are based on thermochemical conversion. In other words, the first three processes rely on the biochemical conversion of more degradable biomass, and the last three focus on adding high-temperature heat to process less degradable woody residues and solid waste. Figure 5.1 shows the possible products, facility types, and technology statuses of these bioenergy processing technologies.

3. Bioenergy Feedstock and Technology

Biomass sources have unique features, including chemical composition, physical status, toxicity, and energy content. These features largely determine how the energy contained in a specific type of biomass can be utilized. However, energy recovery efficiency, economic competitiveness and market opportunity also play a key role in the decision of whether and how a particular biomass processing technology should be used.

For example, wood may be directly used as firewood or fuel for domestic space heating or cooking, but it can also be used as a biomass feedstock for chemical conversion to another final energy product, such as power, heat, biogas, chemical feedstocks, hydrogen, biochar, and soil amendments using biomass processing technology.

Municipal solid waste (MSW) can also be used as fuel in the combustion boiler for power generation, but the consideration of environmental and climate sustainability may decide to first recycle or reuse the MSW and then use either gasification or pyrolysis technology to process the remaining MSW that cannot be recycled or reused to produce more valuable and useful biofuels for transportation,

Type	Technology	Possible Products	Facility Type	Technology Status
Biochemical	Fermentation (enzymatic hydrolysis)	Ethanol, lactic acid, carbon dioxide, hydrogen, oil	Biofuels, agricultural, food, beverage sectors	Mature with efforts to reduce the carbon footprint
	Anaerobic Digestion	Biogas, digestate, soil amendments, fertilizers, and other coproducts, incl. animal bedding	Dairies, food processors, confined animal feedlots, wastewater treatment facilities	Mature with continuing R&D on co-products and high solids/strength digesters
	Transesterification (alcoholysis)	Biodiesel, cellulosic ethanol, chemical feedstocks, hydrogen, other coproducts	Biofuels and biorefineries, especially in the forest products sector	R&D with the pilot and commercial-scale demonstration projects in development
Thermochemical	Waste Combustion (1000°C, fully oxidized)	Power, heat, soil amendments, other co-products	MSW, forest, agricultural, food sectors	Mature, widespread in the developed countries
	Biomass Gasification (700-800°C, oxygen restrained)	Power, heat, biogas, chemical feedstocks, hydrogen, biochar, soil amendments	MSW, forest, agricultural, food sectors	Demonstration emerging into commercialization
	Biomass Pyrolysis (400-700°C, fully oxygen-deprived)	Power, heat, bio-oil, biogas, biochar, chemical feedstocks, soil amendments	MSW, forest, agricultural, food sectors	Demonstration

Fig. 5.1 Bioenergy processing technologies

as well as power and heat. Therefore, the most suitable biomass processing technologies must be selected and adopted by taking these factors into consideration. Figure 5.2 shows the biomass feedstock, processing technology, energy product, energy use, and energy demand.

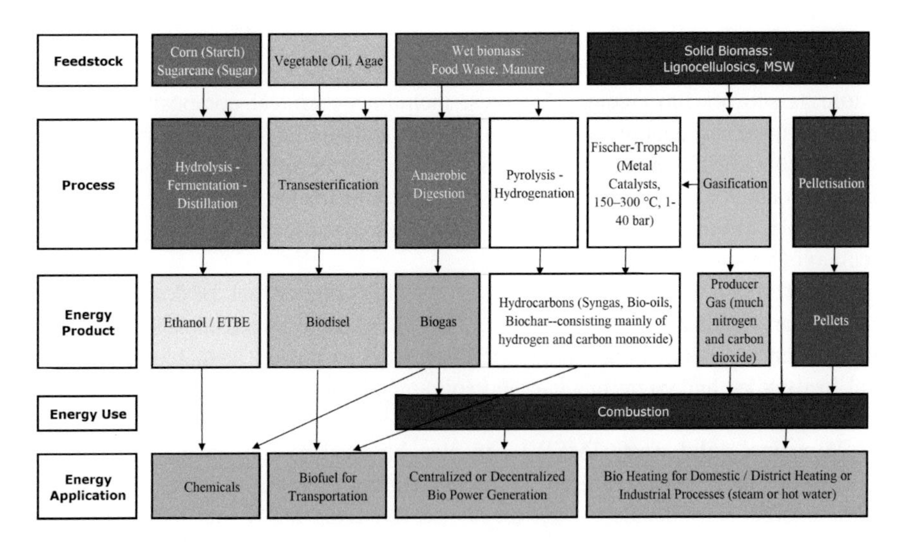

Fig. 5.2 Bioenergy: feedstocks, processes, products and applications

These technologies can be used in combination with pretreatment (such as using organic or inorganic compounds to disrupt biomass structure) and posttreatment (such as stabilization, drying, upgrading, and refining) to produce the final products.

Currently, woody biomass direct combustion or co-combustion with coal is still the most developed biomass sector in the world. However, transportation biofuels harnessed from lignocellulosic residues from agriculture and forestry, as well as biogas or bio syngas produced via biochemical or thermochemical processes of wet and dry biowastes, are emerging sectors.

4. Incineration, Gasification and Pyrolysis of Waste for Energy

Converting organic wastes into biomass energy or products has significant environmental, health, and economic benefits. It can remove enormous amounts of wastes and pollutants from the environment every year, i.e., tens of millions of ton of chemical oxygen demand (COD) emissions, two billion ton of CO_2 emissions, two million ton of NOx emissions, 5.6 million ton of SO_2 emissions, and 300 million ton of dust. Removing these environmental pollutants contributes not only to the sustainability of ecosystems but also to the protection of human health. In addition, the energy recovered from organic wastes can also reduce the production and consumption of fossil fuels.

Incineration, gasification, and pyrolysis of biomass, including organic waste, are technologies used to decompose organic material at various high temperatures and to convert it to energy in other forms of bioenergy, including power, heat, liquid biofuels, or biogas. The major differences among the three technologies include the degree of operating temperatures, the presence and/or amount of oxygen in the heating process, and the type of intended bioenergy products.

Although both waste incineration and waste gasification are usually designed to use high-temperature heat (1000 °C for waste incineration and 700–800 °C for waste gasification) to process waste, the two technologies have significant differences in both how heat processes and what final products the heating processes are intended. The difference in the heating process lies in whether oxygen is involved in the process.

Waste *incineration* is a full oxidative combustion process, meaning that organic waste is directly and completely burned by rapid flame oxidation as a *fuel*, similar to other fossil fuels, leaving nothing but gases and ash. In other words, incineration uses a burner or heat generator.

In contrast, biomass and waste gasification is not designed to use organic waste as fuel but as an organic *feedstock* for chemical conversion to bioenergy. The gasifier is, therefore, a chemical reactor that strictly controls the amount of oxygen used to prevent it from developing to a flame to burn (oxidate) the organic waste. The tiny fraction of oxygen used in waste gasification enhances the efficiency of the high-temperature gasification process. The absence of a sufficient amount of oxygen prevents solid waste from being burned as a fuel but generates high-temperature heat to induce chemical conversion.

The difference in the final products of the two technologies is significant. While the designed output of waste incineration is high-temperature heat used to heat water to produce steam to drive the steam turbine for power generation, the final product of gasification is synthetic gas (syngas). Syngas may be used directly in internal combustion engines or to make products that substitute for natural gas, chemicals, fertilizers, transportation fuels and hydrogen. Pollutants are removed from syngas before it is combusted, so it does not produce the elevated levels of emissions associated with other combustion technologies.

Like gasification, *pyrolysis* also turns solid biomass such as municipal solid waste into other forms of bioenergy. Pyrolysis differs from gasification in the use of lower operating heat at temperatures between 400 and 700 °C, the complete absence of air/oxygen in the heating process, and more transportation biofuels in addition to syngas.

The Fischer–Tropsch (FT) process is the catalytic chemical conversion of syngas (a mixture of carbon monoxide and hydrogen) harnessed from gasification into liquid hydrocarbons, i.e., transportation biofuels such as biodiesel and jet fuel. The process uses metal catalysts (such as metals cobalt, iron, and ruthenium), moderate temperatures between 150 and 300 °C (302–572 °F), and pressures between 1 and 40 bar.

5. Third-Generation Bioenergy

Third-Generation Bioenergy includes nonfood-based biofuels because they are grown as and extracted from nonfood plants such as microalgae and cellulosic biomass such as perennial grasses and fast-growing willow.

Algae fuel (also called algal biofuel or algal oil) is a more advanced form of modern bioenergy technology. It uses algae to produce biofuel as a potentially more sustainable and effective alternative to petroleum-based transportation fuels than the existing first-generation biofuels.

There are two major types of water-bound algae—macroalgae and microalgae. The former is food based, and the latter is nonfood based. Algae fuel is mainly derived from the latter, microalgae, a diverse group of single-celled organisms such as dinoflagellates and diatoms that grow either independently or in colonies. As a nonfood third-generation biomass technology, algae fuel technology has a major advantage, i.e., no competition with food production for land use. Microalgae can be cultivated in open ponds, closed ponds, or translucent photoreactors.

Open-pond algae cultivation is a form of algae farming that involves growing algae in large, shallow, open ponds. Open ponds (i.e., ponds in an open system in open air) are simple and inexpensive open-loop systems for algae cultivation, but they are less efficient than other systems. This method of cultivation is the most common type of algae farming and is used to produce a variety of products, including biofuel, animal feed, and nutritional supplements. Open ponds can be constructed in many different ways, but they all have the same basic components: a shallow pond lined with a waterproof material, an inlet and outlet system for water, and a system of pumps and aerators to keep the water moving and oxygenated. To grow the algae, the ponds are filled with a nutrient-rich solution and then exposed to sunlight. Open systems need to constantly add new water and drain the old water to purify water, such as a natural pond fed by a stream. The algae then grow and reproduce rapidly, forming a thick, green "soup." There are also concerns that other organisms can contaminate the pond and potentially damage or kill the microalgae in the pond.

Closed-pond algae cultivation is a form of algae farming that involves growing algae in a closed system. This method of cultivation is used to produce a variety of products, including biofuel, animal feed, and nutritional supplements. Closed ponds are usually constructed from a variety of materials, including concrete, plastic, and steel. They are designed to contain a nutrient-rich solution and to be exposed to light. The algae then grow and reproduce rapidly, forming a thick, green "soup." The closed system allows for more control over the environment and eliminates the need for pumps and aerators to keep the water moving and oxygenated. It also reduces the risk of contamination from outside sources.

Photobioreactors are the most advanced closed systems, which may be directly connected to carbon dioxide sources (such as dilute CO_2 from the flue gas of fossil-fuel power plants and industrial high-purity CO_2 emitted by SMR, ammonia, and ethanol plants) and use the gas. They are much more efficient in terms of yield and control, but they are much more capital-intensive and costly to operate.

The concept of algae fuel was initially developed because of the oil crisis in the 1970s. The focus on R&D of algae fuels has intensified in the recent climate action and renewable energy transition because algae efficiently use CO_2 and account for more than 40% of the global carbon sink, of which the majority was contributed by marine microalgae. Algae grow very rapidly, with a doubling rate between 6 and 12 h.

As third-generation bioenergy, algae fuel production has many advantages over first-generation bioenergy, such as a high growth rate, reasonable growth density, high oil content, and no need for potable (drinkable) water for algae cultivation, as well as a much higher oil yield and oil content of certain types of microalgae (10–300 times more than other biofuel sources) that can be distilled into biofuels as substitutes for gasoline, diesel, and jet fuel (see Fig. 5.3).

Moreover, with a lifecycle CO_2 footprint of 5.87 pounds (2.66 kg upon combustion and 0.4 kg when driving one mile), the CO_2 footprint of gallon algae fuel is much lower than that of gallon oil (8.89×10^{-3} metric ton) and that of gallon diesel (10.180×10^{-3} metric ton). Therefore, algae cultivation is truly carbon neutral and friendly to the environment because it absorbs enormous amounts of CO_2 during its fast growth and can be cultivated in wastewater or water near power plants or factories, and its spill has no significant or long-lasting adverse effects on the ecosystem.

Algae fuel technology consists of several production stages from lab strain identification and genetic modification to cultivation in the most favorable and sustainable cultivation environments and extraction of algae oil to marketization. Growing rapidly in a wide range of aquatic environments from freshwater to

Crop	Oil yield (gallons/acre)	Oil content (% of dry weight)
Corn	18	3-5*
Soybeans	48	8.1-27.9*
Canola	127	30.6-48.3*
Jatropha	202	38.7-45.8*
Coconut	287	60-70*
Oil Palm	636	29.8-70.3*
Microalgae	**6283-14641**	**20-77**
Schizochytrium sp.		50-77
Botryococcus braunii		25-75
Nannochloropsis sp.		31-68
Neochloris oleoabundans		45-47
Nitzscia sp.		45-47
Chlorella sp.		28-32
Phaeodactylum tricornutum		20-30
Crypthecodinium cohnii		20

Fig. 5.3 Microalgae versus other crops: oil yield and oil content. *Source* Chisti (2007) and * others

saturated saline by taking in CO_2 and releasing oxygen, algae fuel is an environmentally friendly alternative to existing fossil fuels such as oil and natural gas for transportation or heating.

In addition, algae can be used as versatile feedstocks for different types of biofuels, such as the transportation fuels biodiesel and ethanol used in cars, other vehicles, and jets; syngas commonly used for power generation, vehicle propulsion, and industrial heating; biogas primarily used for power generation, heating, and cooking; and other coproducts, such as fertilizers, soap products, and industrial cleaners. Figure 5.4 shows the biomass components of algae, the processes related to each of these components, and the energy forms we can produce using these processes.

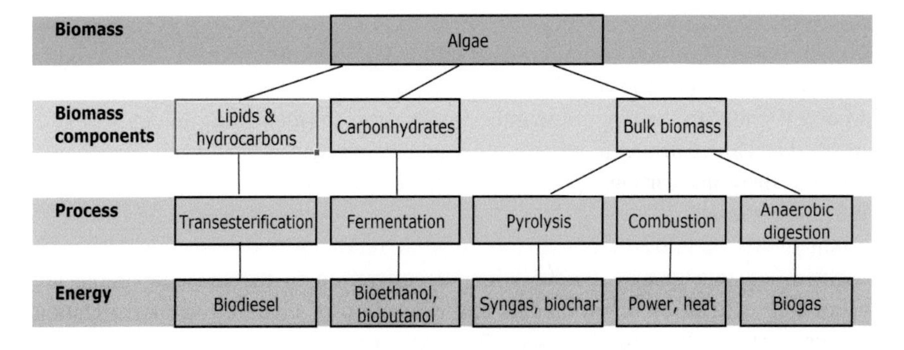

Fig. 5.4 Feedstocks, processes and energy forms of algae

Switchgrass and other types of cellulosic biomass can be grown in areas with highly erodible soils that have marginal food growing quality. In general, third-generation bioenergy avoids competing with fertile farmland most appropriate for agricultural production. The potential for perennial biomass in the future is very promising. This type of biomass production is much more sustainable than annual crop production and can lead to increased yields, improved soil health, and reduced water use. It is also capable of providing many other benefits, such as improved biodiversity, soil carbon sequestration, reduced air and water pollution, and increased resilience to climate change. Additionally, perennial biomass crops provide multiple products and uses, such as food, fuel, fiber, animal feed, and decorative products. Overall, perennial biomass is both viable and sustainable and has enormous potential to become a major source of renewable energy in the future.

Benefits

All modern bioenergy technologies have carbon reduction benefits compared with fossil fuels. At the same time, based on their distinct available organic feedstocks, energy conversion processes, and energy generation costs, each generation of bioenergy technologies presents distinct advantages in terms of feedstock utilization, energy efficiency, environmental impact, and potential applications. These benefits contribute to the broader goal of transitioning to more sustainable and renewable energy sources.

1. First Generation Bioenergy Techniques and Processes

First-generation bioenergy techniques capitalize on mature and well-established agricultural practices for planting and processing biomass. This technical advantage translates to lower costs for energy production and conversion because these techniques are widely understood and implemented, making the adoption of these techniques and processes swift and efficient.

First generation bioenergy utilizes readily available food-based feedstocks, such as corn, sugarcane, and vegetable oils. By repurposing these materials, first-generation bioenergy contributes to the humankind's initial search for alternative solutions for low carbon energy generation and consumption. The conversion of food-based feedstocks into biofuels and bioenergy provides an alternative source of energy. This helps diversify the energy mix and reduce reliance on finite fossil fuels, contributing to greater energy security and stability.

While there are carbon emissions associated with burning biofuels, the overall carbon footprint is relatively lower compared to fossil fuels. This is because the carbon dioxide released during combustion is roughly equivalent to the carbon absorbed by the plants during their growth, resulting in a closed carbon cycle.

The cultivation of food-based feedstocks for bioenergy creates economic opportunities, particularly in rural areas. It supports farmers, processors, and other actors along the supply chain, fostering job creation and rural development.

The use of domestic food-based biomass reduces a country's reliance on fossil fuel imports. This contributes to enhanced energy security, reducing vulnerability to geopolitical and market uncertainties.

By repurposing agricultural residues and byproducts for bioenergy production, first-generation technologies align with principles of environmental stewardship. They mitigate the environmental impact of waste disposal and promote more responsible resource management.

The mature and well-understood nature of these technologies, combined with their ability to utilize food-based feedstocks effectively, makes first-generation bioenergy a practical and accessible step towards sustainable energy production and waste reduction.

2. Second Generation Bioenergy Technologies

Second-generation bioenergy technologies expand the range of feedstocks to non-food crops, agricultural residues, and woody biomass, as well as municipal solid waste (MSW). This broader range of feedstocks reduces competition with food production and enables the utilization of diverse and abundant resources.

Second-generation technologies play a crucial role in reducing waste generated from food production processes. By converting agricultural and forestry byproducts and residues into biofuels, these technologies help minimize the environmental impact of food and forest waste and contribute to a more circular economy. Second-generation technologies address concerns related to deforestation and competition for food resources by utilizing agricultural residues, non-food crops, and waste materials like MSW. This approach minimizes the environmental impact on food production.

These technologies are designed for higher energy conversion efficiencies, allowing them to extract more energy from the same amount of biomass, including MSW, compared to first-generation technologies. Waste-to-energy technologies, such as incineration, gasification, and anaerobic digestion, play a significant role within the second generation. By converting MSW into energy, these technologies offer effective waste management solutions while simultaneously generating electricity or heat.

Due to the increased efficiency and more sustainable feedstock choices, second-generation technologies, including MSW-based WTE, contribute to reduced overall greenhouse gas emissions compared to fossil fuels.

3. Third Generation Bioenergy Technologies

Third-generation technologies focus on novel approaches like algae-based biofuels and microbial fuel cells. These methods promise higher energy yields and lower environmental impact compared to previous generations.

Algae-based biofuels, for instance, also address the land-use concerns of first-generation biofuels and can be cultivated in ponds or specialized systems, eliminating concerns about competing with arable land used for food crops.

Most third-generation technologies, such as algae, can capture carbon dioxide during their growth phase, making them potentially carbon-negative when the entire lifecycle is considered.

Third-generation technologies can produce a range of biofuels, biochemicals, and other valuable products, contributing to the bioeconomy and reducing reliance on petroleum-based products.

Deployment

Bioenergy, the utilization of biomass as an energy source, has a historical development that can be divided into two distinct categories: traditional biomass and modern bioenergy.

Traditional biomass has been used for centuries and relies on basic methods of combustion without significant technological interventions. In ancient times, humans discovered the benefits of burning wood and biomass for heating, cooking, and other energy needs. Throughout history, traditional biomass, including wood, crop residues, and animal waste, has remained a primary source of energy, especially in rural and agricultural communities. However, the use of traditional biomass is often associated with inefficient combustion, indoor air pollution, unsustainable extraction practices, and related ecosystem depletion.

Bioenergy has undergone significant advancements in modern times, with the development and deployment of various technologies across different regions. The first notable advancement in modern bioenergy came with the invention of the gasification process in the early 1900s. Gasification technology, which converts biomass into a gaseous fuel known as syngas, was initially used for small-scale applications such as cooking and lighting in rural areas.

During the mid-1900s, the development of anaerobic digestion technology gained traction. Anaerobic digestion, a biological process that breaks down organic matter to produce biogas (methane and carbon dioxide), started being deployed for the treatment of organic waste and the production of renewable energy. Countries such as Germany, Sweden, and Denmark were pioneers in adopting anaerobic digestion systems for waste management and energy production.

In the 1970s, in response to oil crises and environmental concerns, research and development efforts intensified. The focus shifted toward liquid biofuels, particularly ethanol and biodiesel, as alternatives to fossil fuels. Brazil emerged as a leader in ethanol production, primarily from sugarcane, while the U.S. and Europe focused on ethanol and biodiesel production from crops such as corn, wheat, and rapeseed.

In the early 2000s, second-generation bioenergy technologies emerged, focusing on nonfood biomass utilization, such as agricultural and forestry residues and dedicated energy crops. Notable leaders in this period include the U.S., Brazil, the European Union (Germany, Sweden, Finland, Denmark), and Canada. They made significant advancements in producing cellulosic ethanol through advanced enzymatic hydrolysis and fermentation processes. These countries played a crucial role in demonstrating the feasibility and scalability of cellulosic ethanol production.

Over the last decade, there has been a notable increase in the deployment of modern bioenergy technologies. Advanced biofuel production capacity has expanded globally, with countries such as the U.S., Brazil, Germany, and China leading the way. The growth rate of the bioenergy market has varied by region and technology.

The global ethanol production capacity reached approximately 110 billion liters per year in 2020, with the U.S. and Brazil accounting for most of the production. Brazil's ethanol industry, driven by sugarcane feedstock, has experienced significant growth, with production exceeding 33 billion liters per year. The U.S. has also expanded its corn-based ethanol production to over 60 billion liters per year.

Biodiesel production has also witnessed substantial growth, particularly in Europe and the United States. In Europe, countries such as Germany and France have been leaders in biodiesel production, utilizing feedstocks such as rapeseed oil and waste oils. The U.S. has increased its biodiesel production capacity to over six billion liters per year, primarily derived from soybean oil.

The deployment of biogas systems for renewable energy generation has expanded globally, driven by supportive policies and waste management initiatives. Germany, with its feed-in tariffs and incentives, has emerged as a leader in biogas production. The country had over 10,000 biogas plants with an installed capacity exceeding 4500 MW in 2020. Other countries, such as Denmark, Sweden, and Austria, have also experienced significant biogas growth.

The market for modern bioenergy has experienced notable expansion in recent years, driven by several factors. Increasing concerns about climate change, fossil fuel depletion, and the need for sustainable energy sources have led governments and industries to prioritize bioenergy development.

Supportive policies, financial incentives, and research initiatives have encouraged investment in advanced bioenergy technologies. Geographic variations in market growth exist, with countries such as Sweden, Finland, Germany, the U.S., and Brazil emerging as leaders in modern bioenergy adoption.

As a renewable energy source, bioenergy plays a key role in the energy transition. In the period between 2010 and 2021, biomass use accounted for 10% of the total world primary energy consumption. Biomass use grew from 13,529 TWh in 2010 to 18,474 TWh in 2021, accounting for an annual average growth rate of 3.3% (see Fig. 5.5).

The largest use of modern bioenergy in industry grew from 2250 TWh in 2010 to 3334 TWh, accounting for a growth rate of 48% or an annual average growth rate of 4%. The second largest use of modern bioenergy, power and heat rose from 1111 in 2010 to 2584 TWh, accounting for the highest growth rate of 133% or an annual average growth rate of 12%.

The biofuels for transportation grew from 667 to 1195 TWh, accounting for a growth rate of 79% or an annual average growth rate of 7%. The biofuel supply dropped sharply in 2020 because of reduced transportation energy demand and

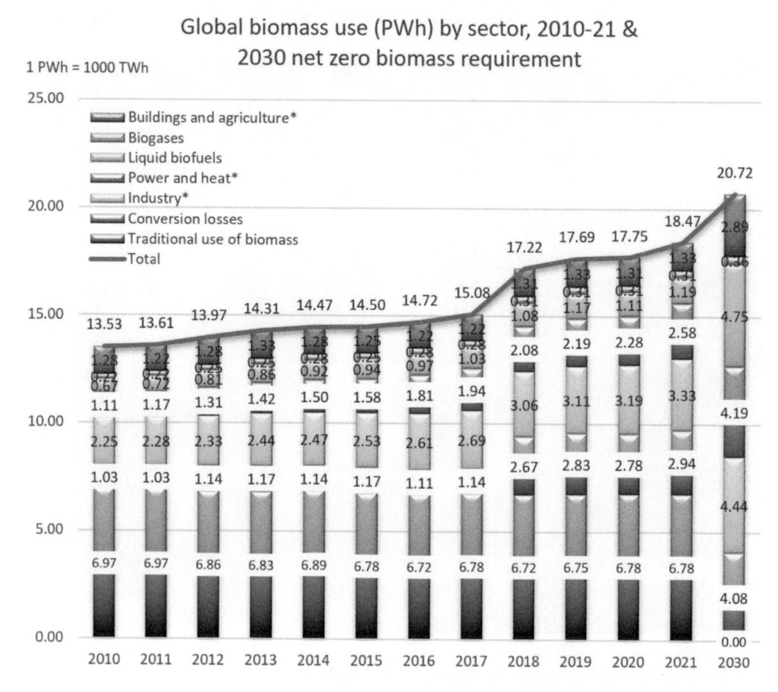

Fig. 5.5 Biomass use composition and net zero requirements by 2030. *Data Source* IEA. CC BY 4.0

restricted blending during the COVID-19 pandemic. In 2021, the biofuel supply recovered to levels close to those of 2019 but was affected by high feedstock prices.

In 2021, modern biomass accounted for 47.4% of the total biomass use, of which 18% was used for industrial production, 14% for power and heat, 7.2% for building and agriculture, 6.5% for liquid biofuels, and 1.7% for biogas. At the same time, traditional biomass still accounted for 36.7%, and conversion loss, i.e., the loss of energy during biomass production and consumption, accounted for 15.9% (see Fig. 5.6).

In the power and heat sectors, bioenergy's contribution has increased 88% overall since 2011. China continued to lead in biopower generation, with an overall growth in biopower generation of 450% since 2011 or an average annual growth rate of 45%. Biopower generation also showed robust growth in some other Asian and European countries. However, the biopower generation of the next three top countries, the U.S., Brazil, and Germany, did not show substantial growth in recent years.

The main biomass source for power generation was solid biomass (83%), which included bagasse (residue of sugar production from sugarcane) and municipal solid waste. The remaining bioenergy power generating capacity included 15% biogas and 2% liquid biofuels.

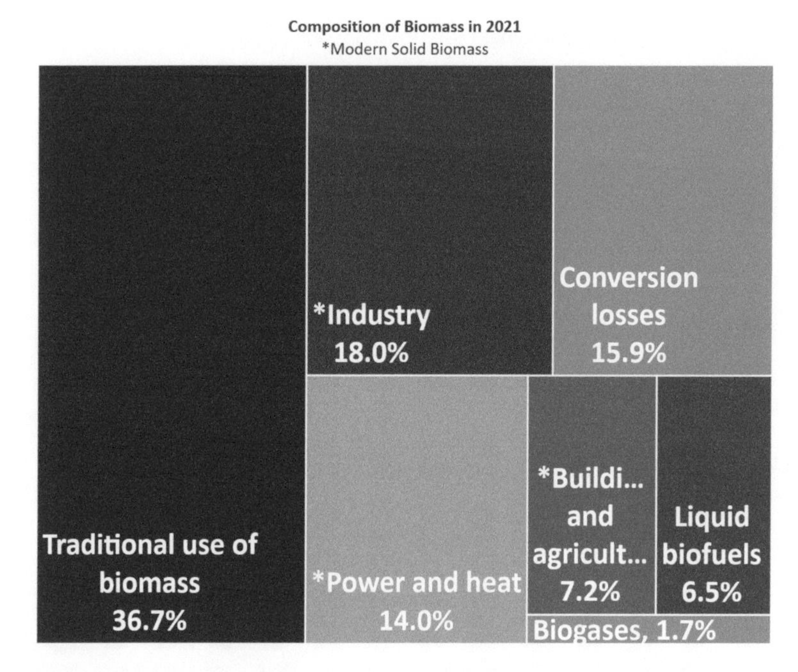

Fig. 5.6 Composition of biomass use. *Data Source* IEA. CC BY 4.0

The global installed biopower capacity reached 151 GW in 2022, accounting for a 5.1% increase compared to that in 2021. This annual growth rate of the installed bioenergy power generating capacity was slower than the average annual growth rate of 8% between 2001 and 2010, 7% between 2010 and 2022, and 6% between 2017 and 2022. This indicates that although the biopower has generally kept rising, the growth rate has gradually reduced over time (see Fig. 5.7).

The global biopower generation using biomass grew from 372 TWh in 2012 to 670 TWh in 2022, accounting for an annual average growth rate of 6% (Fig. 5.8).

As part of the biopower sector, the contribution of waste-to-power plants (also called waste-to-energy, trash-to-energy, municipal-waste-incineration, energy-recovery, or resource-recovery plants) to power generation is still limited. These plants are waste management and power generation combined facilities based on waste combustion technology. Today, there is more than 2600 waste to power plants around the world. Sweden is the world's leader in using waste combustion for power technology. It recycles nearly half of its total municipal solid waste, burns the other half, and sends only 1% of its MSW to landfills.

Waste-to-power technology is a mature technology that carries out both waste management and power generation. The operation of waste-to-power plants is also more profitable than that of fossil fuel-fired power plants because waste-to-power plants do not have to pay for the solid waste they use and receive revenue for receiving solid waste instead of damping it to landfills in the U.S. This bonus is called the tipping fee, which was estimated to be $54 per ton in 2021. This fuel

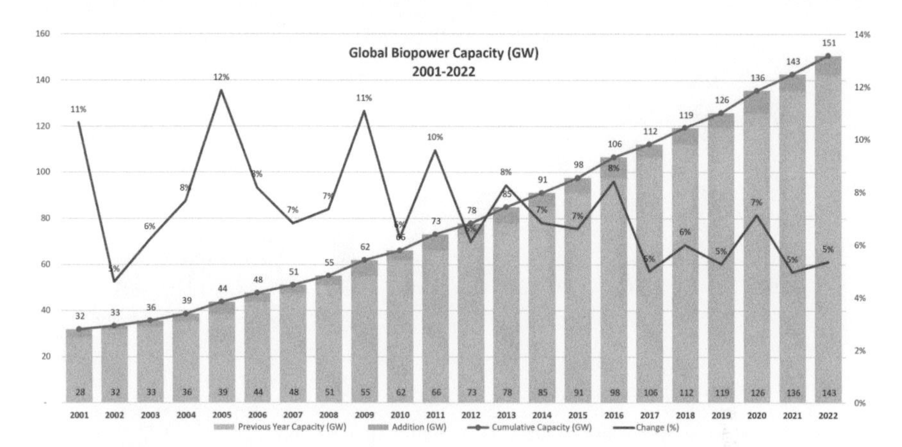

Fig. 5.7 Global biopower capacity (2012–2022). *Data Source* IRENA, 2023

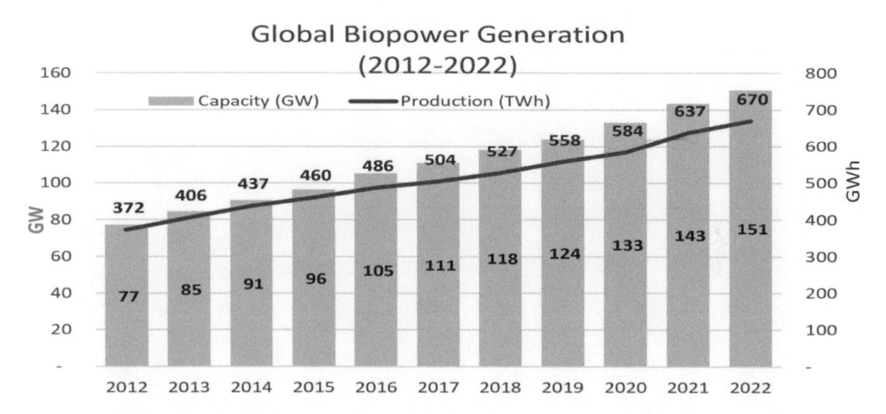

Fig. 5.8 Global biopower capacity and production between 2012 and 2022. *Data Source* IRENA, 2023

cost savings for waste-to-power plants was substantial because it might make up 45% of the power generation cost for a coal-fired power plant and 75% or more of that for a natural gas-fired power plant. Figure 5.9 shows the share of MSW combusted for power generation in Sweden and in the other nine top countries.

In 2022, global biofuel production recovered from the sharp decline in 2020 caused by the COVID pandemic and reached 1187 TWh. The U.S. led the biofuel market with a dominant share of 38% of the market and 452 TWh of biofuel production. The other two top market leaders were Brazil and Indonesia, with 21% and 9% market shares and 254 TWh and 108 TWh of biofuel production, respectively. China and Germany each had a 3% market share, whereas Argentina, India, Netherlands, Thailand, and France each accounted for a 3% market share (Fig. 5.10).

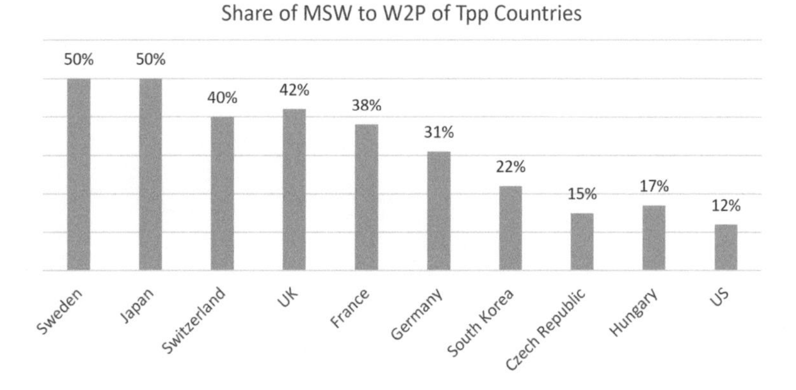

Fig. 5.9 Top 10 waste-to-power countries. *Data Source* IEA. CC BY 4.0

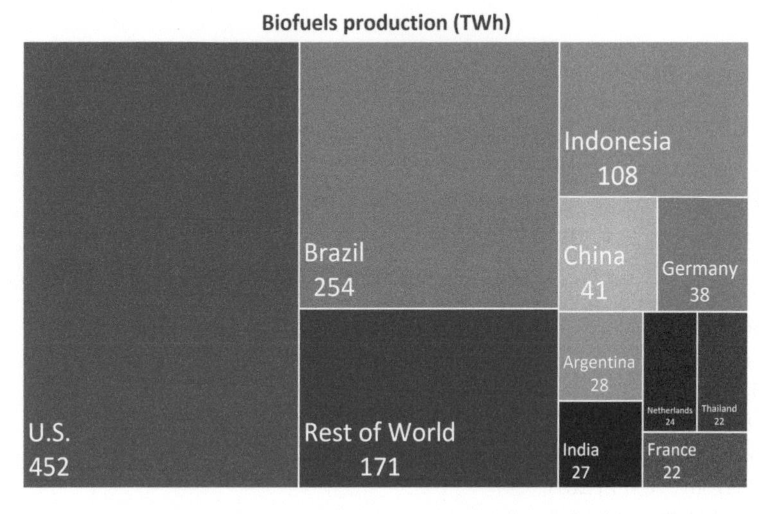

Fig. 5.10 Global biofuel production 2022. *Data Source* Our World in Data, 2023

China led in world biogas production with the highest number of biogas plants (over 100,000), a large number of household biogas units, and the highest annual biogas production (72,000 TWh). Germany ranks second in biogas production, with over 10,000 operating digesters and 120 TWh of annual biogas production. The U.S. has over 2300 biogas sites, of which 2000 generate power: 1269 water resource recovery facilities using an anaerobic digester, 652 landfill gas projects, 316 anaerobic digesters on farms, and 66 stand-alone systems that digest food waste.

Most biogas-producing countries use biogas for heat and power generation. Sweden was an exception by using more than 50% of its biogas as transportation fuel. Germany was second in absolute numbers in terms of using biogas as a fuel.

Both countries have had the largest markets for biomethane in recent years, but a growing interest in biofuels is seen in other countries as well. The UK has now taken over the second position from Sweden, using increasingly more biomethane for heat and electricity production but also as vehicle fuel.

Currently, microalgae cultivation is more actively researched and developed than macroalgae as a nonfood-based third-generation biomass technology and a potentially much more energy-rich and sustainable alternative to existing biofuel sources, such as corn and sugarcane. The high lipid content, high growth rate and ability to rapidly improve strains and produce coproducts without competing for arable land make algae fuel an exciting addition to the sustainable fuel portfolio. Approximately 500 algae and spirulina production plants were in operation across 23 European countries. In 2020, more than 50% of these companies, with Germany, Spain, and Italy as the top three countries, produce microalgae. According to the International Energy Agency (IEA)'s *2019 Global Status Report*, the global production capacity of algae fuels was estimated to be approximately 70,000 ton in 2017. The majority of this production came from three large-scale facilities in South Korea, Germany, and France. Additionally, there were numerous small-scale projects in countries such as Australia, Brazil, India, and the U.S. In 2018, the production capacity of algae fuels was estimated to be 140,000 ton.

Challenges

A. Net Zero Requirement Challenge

Although bioenergy supply grew at an annual average rate of 10% in the past decade and had an increasing share in the global energy supply in the last two decades, this share only increased by 1% from 5.7% in 2010 to 6.7% in 2021. This slow relative expansion was caused by two factors—the growing global energy demand and the faster relative expansion of solar, wind, geothermal and marine power (from 3.2% in 2010 to 5.2% in 2021). However, the Net Zero target by 2050 requires bioenergy to reach 17.5% by 2030, which means that bioenergy supply must grow at an annual rate of 18% on average by 2030, a much faster rate than the rate of 10% in the past decade (see Fig. 5.11).

The supply of bioenergy for the transportation sector, i.e., liquid biofuels, needs to quadruple from 1248 TWh in 2021 to 4962 TWh in 2030, and that of bio jet kerosene for air travel must increase to account for more than 7% of all aviation fuel use in 2030. The bioenergy used to replace fossil fuel in industry (mostly in cement, pulp and paper, and light industry) must rise from 3055 TWh in 2021 to over 4722 TWh in 2030. Bioenergy used for power generation needs to increase from 750 TWh (approximately 2.5% of total demand) in 2021 to approximately 1350 TWh (approximately 3.5% of total demand) in 2030.

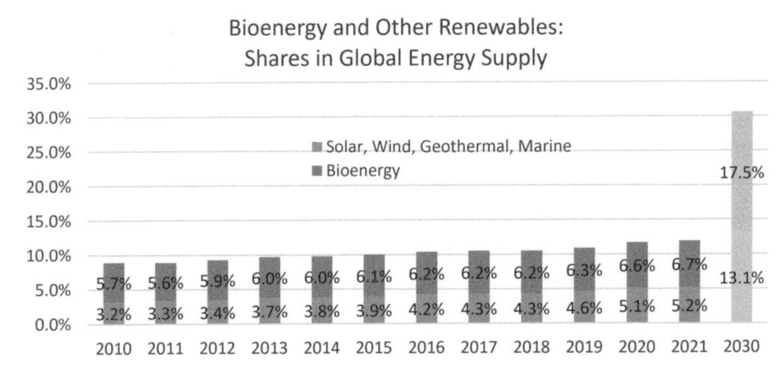

Fig. 5.11 Bioenergy and other renewable energy contribution. *Data Source* IEA, IRENA

B. Technological Challenges

1. Traditional Biomass Challenges

Although considerable progress has already been made in modern bioenergy—solid biofuels, biogas, biodiesel, and liquid biofuels – traditional biomass still dominates and causes many challenges. The term traditional biomass refers to direct burning biomass, such as wood, agricultural residue, and animal dung.

Today, nearly 50% of biomass is used by direct combustion for cooking and heating, mainly in developing countries. Traditional biomass wood, either burned directly or burned together with coal, remains the dominant biomass use in the world. In addition, many low-income countries and regions produce substantial amounts of crop straw, livestock and poultry manure, forestry residues, agricultural processing residues, and domestic waste. Most of these "waste" biomass resources are not effectively used but are burned in the open air or randomly damped and become sources of air, water, and soil pollution, along with fossil fuels. In 2021, more than 35% of the total global biomass consumption was biomass use with traditional cooking and heating methods. Globally, approximately 2.4 billion people (or a third of the world population) cook using open fires or inefficient stoves fueled by biomass (wood, animal dung and crop waste) with or without coal.

Directly burning biomass has multiple issues. First, burning biomass as fuel is a major contributor to air pollution and climate change, similar to burning fossil fuels. Second, traditional biomass provides extremely poor energy quality. The energy loss during the combustion of this traditional biomass for cooking, for example, can be as high as 90%. Third, the direct combustion of agricultural and forestry residues produces higher hazard gases and smoke that severely pollute indoor and outdoor air and cause respiratory diseases in residents. The pollution of traditional biomass is linked to five million premature deaths in 2021 alone. Fourth, the reliance on traditional biomass in poor regions of developing countries to meet the increasing energy demand reduces the availability of traditional biomass and causes environmental and ecosystem degradation.

Although traditional biomass consists of renewable energy sources, all traditional biomass consumption practices of directly burning it are wasteful, unsustainable, and harmful.

2. Biomass Supply and Supply Chain Challenges

There are unique challenges to various biomass energy technologies. One of the main challenges to large-scale biomass expansion is biomass energy density. If the moisture of a biomass feedstock is 35%, every ton of biomass transported and processed only has 65% biomass. Depending on the processing technology used, feedstocks might need pretreatment, such as dehydration, i.e., removing 35% of moisture before further processing. In addition, the shapes of biomass feedstocks, such as chipped, pelletized, rounded, and baled wood, also strongly impact the feedstocks' bulk density and affect their transportation costs.

The unique challenges of the biomass sector like this run through its entire supply chain from biomass species selection, cultivation, harvesting/collection, storage, transportation, pretreatment, and biomass processing at the biomass energy plant. In addition to the low density of biomass affecting transportation and processing costs, these challenges include seasonal variation affecting the regional and seasonal availability of biomass, cost of biomass feedstock, and storage issues; geographic dispersion affecting transport costs; and maturity of the biomass processing industrial and supply chain affecting long-term contracts and technological developments.

3. Food-Based Biofuel Challenges

Food-based bioenergy, i.e., first-generation biofuels such as ethanol harnessed from corn and other crops, competes with food production for land use. They face increasing social and environmental challenges, which significantly affect their further expansion.

In 2021, over two-fifths (3293 million people) lived below $5.50 per day. Malnutrition, hunger, repeated infection, and inadequate psychosocial stimulation remain serious social issues that cause one out of four children in the world to suffer from stunting, i.e., impaired growth and development. In addition, population growth and increasing incomes in developing countries will cause the food demand in the world to sharply rise in the future.

Biofuel feedstocks include substantial amounts of crops that would otherwise be used for human consumption directly or indirectly as animal feed. Converting these agricultural crops into biofuels may divert more agricultural or forestland areas to bioenergy production, increase the use of polluting inputs such as fertilizers, increase food prices, and add unintended adverse impacts on the human consumption of food and the environment. The expansion of food-based bioenergy

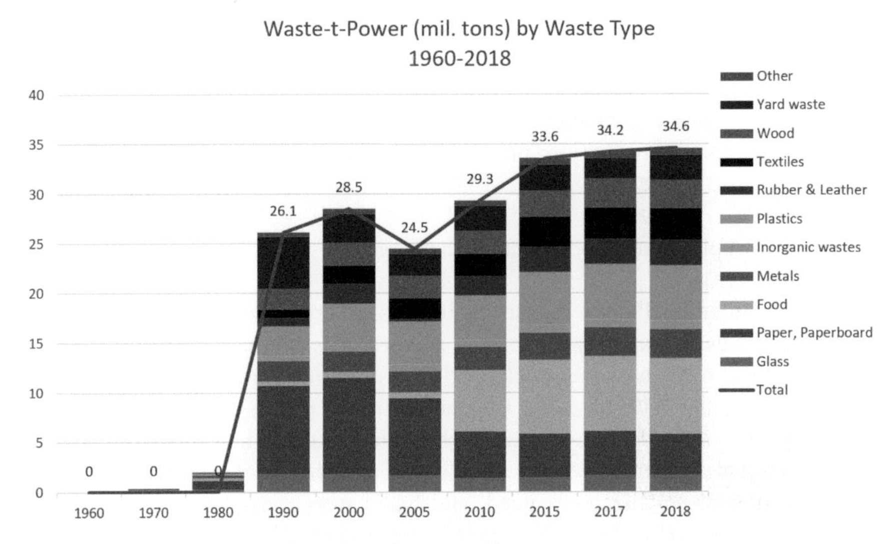

Fig. 5.12 Waste-to-power 1960–2018. *Data Source* U.S. EPA

that relies on traditional agriculture will further aggravate the skyrocketing food demand in the world and lead to catastrophic ecosystem failures.

4. Waste-to-Power Challenges

The waste incineration technology that dominates the current waste-to-power sector in developed countries causes multiple issues, and the use of this technology has become increasingly controversial and is considered outdated today. The historical data on waste to power during the period between 1960 and 2018 published by the U.S. Environmental Protection (EPA) illustrate some of these issues associated with the use of waste-to-power technology (see Fig. 5.12).

The total waste (34.6 million ton) used to generate power in the U.S. in 2018 was broken down into 11 types. More than forty percent of this waste could have been recycled. For example, paper and paperboards (20%), textiles (9%), metals (9%), and glass (5%) were 100% recyclable. If these waste materials had been recycled, the total amount of combusted waste would have been reduced by at least 40%. At a modest cost of recycling, the recovered value of recycled materials would far exceed that of power generated from combusted materials.

Recycling these materials would also have avoided the cost and environmental impact of additional input (such as capital, land, fertilizer, energy, and labor) and a lengthy process (such as growing trees and cotton or extracting minerals and manufacturing products) for their replacement.

Other waste types can either be composted or processed in a more environmentally friendly manner than simply burned as fuel for power generation. Waste food (22%) and yard waste (8%) can be either converted into fuels using more controlled and environmentally friendly approaches or composted (see Fig. 5.13).

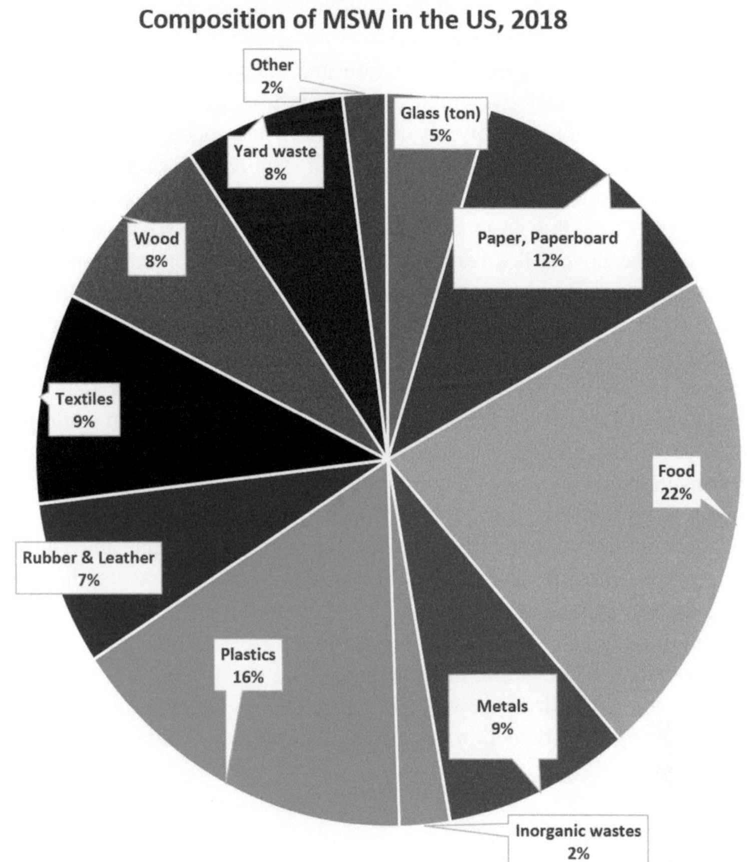

Fig. 5.13 U.S. waste to power generation by waste type, 2018. *Data Source* U.S. EPA

Waste-to-power plants cause less air pollution than coal-fired power plants but more than natural gas-fired power plants. In addition, although waste-to-power plants emit significantly less carbon and methane into the air than landfilled waste, burning toxic municipal waste emits toxic chlorinated byproducts such as dioxins, which are often more harmful to the environment and human health than toxic waste before combustion. In addition, waste combustion also releases odor pollution and fly ash that contain toxic metals such as lead, cadmium, copper, and zinc as well as insignificant amounts of dioxins and furans, which pose a potential health hazard.

Low- and middle-income countries face even more challenges in their waste-to-power projects. The greatest challenges are the lack of funds to afford high capital costs and the lack of technologies to meet air pollution control requirements. Most of these countries cannot afford waste-to-power technology despite the savings from free solid waste "fuel" and avoided landfilling costs. These countries have significant technological barriers in the deployment of advanced waste incineration and power generation technologies, such as advanced boiler systems, supporting auxiliary equipment technology, deposition and slagging of combustion devices and anti-corrosion technology, which need breakthroughs.

Biomass gasification and pyrolysis are more advanced and complex solid biomass and waste processing technologies that require high investment and operational costs. In addition, additional costs are involved in adding an air purification system to further treat flue gases from gasification and pyrolysis. Based on the concentrations in the flow, ashes resulting from the pyrolysis process contain a high heavy metal content. Pyrolysis plants usually produce some oil and water if there is significant condensation. It also generates a small amount of tar, which is polluting unless destined for appropriate usages such as asphalt or roof shingles. Although gasification and pyrolysis are generally considered more efficient than incineration, these advanced biomass technologies also face objections from environmental groups. One of the critical issues that is often debated is whether these waste energy conversion technologies require more energy than the waste produces. This issue needs to be addressed and clarified through more solid research.

Low- and middle-income countries have even more financial and technological barriers to gasification and pyrolysis waste-to-power processes, although these technologies are capable of delivering cleaner processes and have much less adverse environmental and health impacts. Gasification R&D faces technological barriers such as small scale, low efficiency, and the difficulty of the byproduct separation process. Co-firing power generation technology has yet to establish a complete co-firing ratio detection system, high-efficiency biomass fuel boiler and feeding system.

5. Challenges to Algae Fuel

To maximize the growth of microalgae in open ponds, a significant water source is required for large-scale cultivation. The best algae growth requires 10–15 h of sunlight a day and temperatures between 16 and 27 °C (60–80°F).

Because these hot temperature levels for the best growth cause water in the open raceway environment to evaporate, algae cultivation for biofuels uses much more water than some other renewable energy sources, such as solar PV and wind technologies. The high water usage of microalgae cultivation is a real concern, especially for regions that already struggle with water shortages, because algae fuel generation will compete with agricultural, industrial, and human water demands.

Large-scale expansion of microalgae production requires substantial amounts of fertilizer. For example, the replacement of 5% of the current U.S. petroleum transportation fuels with algae fuel would require 15% of all the fertilizer produced in a year.

The increased fertilizer use causes environmental and climate concerns because it would aggregate the problems already associated with it in agriculture, i.e., run-off may pollute water sources, creating algae blooms that can kill aquarian and

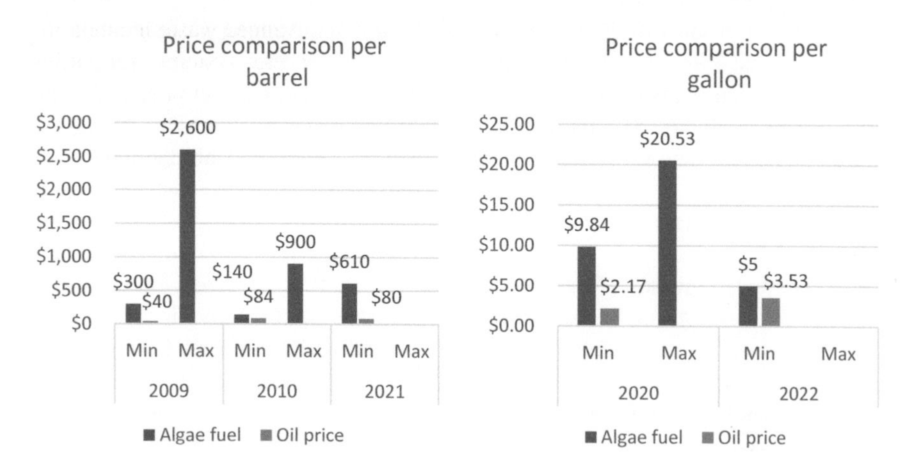

Fig. 5.14 Price comparison of algae fuel and oil by barrel and gallon

marine wildlife. In addition, fertilizer production and consumption require energy and emit CO_2, negatively impacting the carbon neutral status of algae biofuels.

The costs of producing algae fuels and the price of using them are still much higher than those of equivalent fossil fuels and still lack market competitiveness.

Figure 5.14 shows a price comparison of algae fuel and fossil oil by barrel and gallon. This indicates that the algae fuel price dropped significantly between 2009 and 2022. The costs of a barrel of algae-based fuel fell from $300 to $2600 in 2009 to $610 in 2021. However, its production costs of $9.84 using the open raceway approach and $20.53 using the closed photobioreactor approach were still much higher than that of $2.17 for gasoline in 2020.

In addition, despite the sharp price reduction in the minimum algae fuel price between 2021 and 2022, i.e., from $9.84 in 2021 to $5 in 2022, the reduced algae fuel price was still more than 40% higher than the minimum gasoline price in 2022. Further R&D and marketization are necessary to further reduce the costs of algae fuel production.

Although algae fuels have exciting achievements and high potential as a viable alternative to fossil fuels, advanced algae fuel technology is still at an early stage of R&D. There remain a large number of technological and economic challenges to algae production from species selection, cultivation, and harvesting to downstream operations for converting microalgal organisms into market-ready products.

The R&D of algae fuels has been rapidly progressing. However, the speed, extent, and required support of algae fuel projects are largely absent. Moreover, most current R&D results are from algae fuel research labs or from small-scale algae fuel projects. Because of the lack of data from large-scale production, it is also difficult to develop a clear overview of the sprawling, dynamic algae fuel industry, its technological advancement, and the environmental benefits and impacts of algae cultivation and algae fuel production. Large-scale experiments, on which the expanded deployment of algae fuel technologies depends, are largely lacking.

Algae fuel production is a complicated and time-consuming process involving multiple steps. First, the most productive algal species must be selected. Then, algae must be cultivated, algae oil must be extracted through an oil press, solvents, or supercritical fluids, and the extracted oil needs to be refined. This process is much longer and therefore more expensive than petroleum fuel production.

6. Cellulosic Biomass Challenges

Certainly, switchgrasses, fast-growing willows, and other types of cellulosic biomass in areas not suitable for crop production have significantly higher potential for sustainable bioenergy than first-generation food-based biomass. However, this potential has not yet been used at a large scale for several reasons. First, developing and implementing long-term large-scale biomass systems is much more expensive than using the existing systems for annual crops because there is a need to invest in longer-term infrastructure and ongoing maintenance, such as intensive tillage, soil improvement, pest control, planting, and harvesting, in areas not suitable for crop production. Second, there is a lack of awareness among policymakers, producers, and society about the environmental and economic benefits that third-generation biomass can offer. This has led to the perception that it is too expensive and therefore not worth investing in and has caused many governments and organizations to stop providing incentives for farmers and landowners to take up perennial biomass production.

7. Political, Institutional, and Socioeconomic Barriers

Currently, there are major challenges to the efficient expansion of sustainable bioenergy. Many countries are facing political, institutional, and socioeconomic barriers to the research, development, and deployment of sustainable bioenergy technologies.

First, government fossil fuel subsidies constitute major obstacles to the innovation and deployment of advanced bioenergy technologies. According to the International Institute for Sustainable Development (IISD), more than 90 countries still subsidize fossil fuels in some way. These subsidies are estimated to total approximately $450 billion a year worldwide. This indicates that governments around the world are still actively supporting the use of fossil fuels.

On the other hand, the lack of government, public, and business actions represents major barriers to the urgently needed acceleration of innovation and deployment of advanced bioenergy technology. These barriers include insufficient public awareness of the benefits of the second- and third-generation bioenergy technologies; lack of clear objectives, strategies, and policy coordination for advanced bioenergy research, development, deployment, and decision making at international, national, and local levels; weal government regulation and policy support in advanced bioenergy technologies; policy uncertainty and institutional red tape in the deployment of these technologies; and lacking information and communication on the objectives and strategies between bioenergy stakeholders.

In summary, the expansion of sustainable bioenergy deployment faces major challenges because the current biomass and bioenergy sector is still largely dominated by traditional biomass, first-generation bioenergy, and outdated incineration-based waste to power. Traditional biomass, i.e., directly burning biomass for heating and cooking, is both inefficient and harmful to the environment and human health. The first-generation bioenergy technologies that convert food-based feedstocks into biofuels directly compete with food production in land use, put undue pressure on human lives and set natural limits on their own future development. Incineration-based waste-to-power technology eliminates the possibility of recycling substantial amounts of recyclable materials and releases harmful toxins and pollutants in the air that pollute the environment and cause human health risks.

Solutions

A. Policy, Regulatory and Socioeconomic Solutions

Governments worldwide play a crucial role in supporting the bioenergy sector's research, development, and deployment through clear visions, targets, and strategies. Financial support from governments can be directed toward bioenergy technologies with long-term environmental benefits and fewer adverse impacts. Concurrently, governments can formulate policies aligned with carbon–neutral objectives under the UN Sustainable Development Goals. These policies facilitate accelerated research, development, and deployment of advanced bioenergy technologies, encompassing solid waste conversion, biogas, biofuels, and algae fuels, which generate cost efficiencies exceeding conventional sources and rivaling renewable fuels.

In Europe, the European Commission's proposal outlines a comprehensive plan, aiming to sustainably derive a growing portion of jet fuel. The EU and specific countries such as Sweden and Finland are transitioning toward full biofuel-based transportation by 2040, guided by directives promoting renewable energy, biofuel usage, and bioenergy sustainability.

The U.S. Inflation Reduction Act introduces tax credits for cleaner aviation fuels. President Biden's commitment to produce three billion gallons of sustainable fuel and reduce aviation emissions by 20% by 2030 further exemplifies targeted efforts. The U.S. has set a 2030 target of 30% biomass energy in transportation fuels, supported by initiatives such as the Renewable Fuel Standard, Biomass Crop Assistance Program, and Biorefinery Assistance Program.

China, Japan, and India also implement robust policies promoting bioenergy, ranging from renewable energy regulations to biofuel incentives, sustainability standards, and research programs. At the international level, governments should collaboratively establish clear long-term bioenergy objectives, prioritize strategies, and ensure coordinated cross-sector efforts to minimize adverse impacts while accelerating bioenergy development.

To gain public and business support, raising awareness of advanced bioenergy's environmental benefits is pivotal. Education, public campaigns, and information-sharing initiatives can effectively engage stakeholders. Additionally, providing accessible information on bioenergy resource availability and locations can aid project developers in identifying optimal feedstocks and project sites. Ultimately, a well-coordinated approach, strategic incentives, and robust awareness campaigns can bolster the bioenergy sector's growth and sustainability on a global scale.

B. Market Solutions of Sustainable Transportation Fuels

The endeavor to decarbonize transportation fuels, particularly in the aviation sector, requires significant investment and market-driven solutions. Key players in the aviation industry, for example, are taking steps to address the challenges of cost and scalability, leading to notable initiatives and strategies that are shaping the landscape of sustainable aviation fuel (SAF) technology adoption.

United Airlines, a prominent consumer of sustainable fuel in the U.S., has introduced, in collaboration with other companies, a $100 million venture capital fund dedicated to sustainable aviation fuel technology. This substantial investment underscores the growing interest in and commitment to advancing SAFs. Concurrently, Boeing's commitment to doubling its utilization of sustainable fuel in the present year emphasizes the industry's collective push for carbon reduction.

Despite these challenges, progress is evident in the utilization of sustainable aviation fuel. United Airlines integrated these fuels, amounting to less than 1% of its total consumption in the previous year. This usage trend is gaining momentum across the sector, with Boeing and other airlines increasingly adopting sustainable aviation fuel to mitigate emissions.

Sustainable aviation fuel currently stems from sources such as used cooking oil and agricultural waste. While production is nascent, a few companies, including World Energy and Neste, have emerged as influential forces, producing sustainable aviation fuel at a scale relevant for major airlines.

Start-up Contributions Gevo, a Denver-based start-up, exemplifies entrepreneurial contributions to sustainable aviation fuel. The company's ethanol-based approach has led to the initiation of a plant in South Dakota, highlighting the role of innovation in diversifying fuel sources.

United Airlines' Sustainable Flight Fund United Airlines' proactive stance extends to the creation of a Sustainable Flight Fund through its in-house venture capital arm. The fund's mandate to invest in sustainable fuel production and associated technologies underlines the industry's commitment to innovation and change.

In concert, these instances illustrate the multifaceted efforts propelling sustainable aviation fuels into the forefront of fuel solutions. The industry's collective pursuit is reinforced by strategic investments, policy imperatives, and emergent technologies. The aviation sector's resolute aspiration to achieve carbon neutrality by 2050 underscores the clear market support and vocal alignment with environmental goals.

To optimize biomass and bioenergy production costs, strategic market support is indispensable. Effective supply chain management for bioenergy projects, rooted in thorough assessment and design, ensures optimized operations. The alignment of the global aviation industry's carbon neutrality ambitions with tangible market support emphasizes the pivotal role of effective policies and collaborative partnerships. The collective adoption of sustainable aviation fuel targets by the U.S. airlines affirms the industry's resolve and propels operational changes.

C. Technological Solutions

1. Traditional Biomass Solution

To solve the ecological, environmental and health impacts, it is important to reduce our reliance on the traditional practice of traditional biomass consumption. This means that we should gradually reduce the direct burning of traditional biomass for cooking and heating. This reduction should be supported by the increased deployment of renewable energy technologies in converting traditional biomass to renewable energy in the future. This is possible because traditional biomass consists of renewable energy sources, such as agricultural, industrial, forestry, and municipal waste. Converting traditional biomass and eliminating its direct burning will resolve the wasteful, unsustainable, and harmful impacts of traditional biomass on human health, the environment, and the climate (Fig. 5.15).

The use of this traditional biomass is set to fall to zero by 2030 in the Net Zero scenario to achieve the *UN Sustainable Development Goal 7 on Affordable and Clean Energy*. Traditional biomass uses will be replaced by modern bioenergy usage, which will nearly double from approximately 42 EJ in 2021 to 80 EJ in 2030.

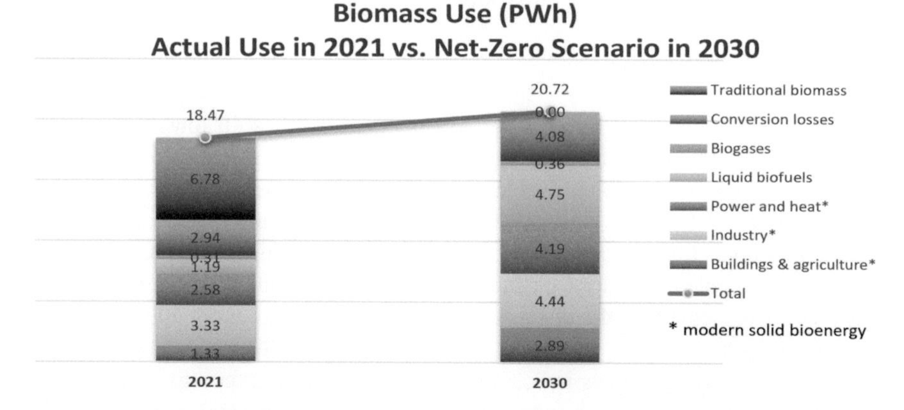

Fig. 5.15 Traditional use of biomass must stop in 2030. *Data Source* IEA, 2021

2. Expanding Second-Generation Bioenergy

To solve the issues of the current food-based first-generation biomass/bioenergy sector competing with food production on land use and causing loss of ecosystem preservation and the homes of indigenous people, R&D in nonfood-based bioenergy, such as food waste, crop residues, forest processing residues, and algae fuels, should be accelerated to achieve breakthroughs in productivity and price reduction, similar to solar PV deployment in recent decades.

Second-generation bioenergy technologies are designed to convert crop residues, animal manures, and forest processing residues that are largely burned directly as traditional biomass into liquid biofuels such as aviation kerosene, biodiesel, and ethanol. To reach this objective, the commercialization of related technologies such as clean pretreatment, transesterification, and anaerobic digestion needs to be accelerated. The research, development and deployment of lignocellulosic technologies for traditional aviation fuel, Fischer–Tropsch synthetic oil, and oil bio aviation fuel also need expansion, acceleration, standardization, and commercialization.

The deployment of second-generation bioenergy technologies should be accelerated to generate power, heat for space heating and hot water supply in buildings, and processes in agriculture and industry, as well as transport fuels, industrial feedstocks, and fertilizer. For example, the development and deployment of biomass-based aviation fuel should be accelerated as one of the most promising second-generation bioenergy options in the near and medium term.

Second-generation bioenergy can become competitive, especially when cheap- or even negative-cost biomass feedstock, such as various waste sources and residues, is used. While enhanced innovation can help a number of bioenergy technologies approach maturity, accelerating further technology R&D can allow promising technologies to reach commercialization and to achieve cost competitiveness and economies of scale.

3. Waste to Energy Solutions

In recent years, several countries have witnessed a notable shift toward more sustainable and efficient waste-to-energy processes, with the adoption of advanced gasification, pyrolysis, and anaerobic digestion systems. Germany, Sweden, Denmark, and the U.S. focus on anaerobic digestion for organic waste processing, producing biogas for heat and power, as well as organic fertilizers. These countries have also integrated advanced gasification technologies into their waste management strategies, enabling the conversion of various waste streams into syngas for energy production.

Their successful practices indicate that advanced waste-to-energy technologies can help further reduce environmental and climate impacts and energy use and go beyond power generation to biofuel generation, and research and development in advanced waste-to-energy technologies should be accelerated and expanded so that these technologies can become truly renewable and sustainable. To reach this objective, the following steps must be implemented.

First, governments should implement more stringent environmental regulations and monitoring systems to ensure that waste is managed and processed responsibly, emissions are minimized, and waste-to-energy plants do not cause damage to the environment or human health. These regulations include setting stricter emissions standards for air pollution, banning incineration of toxic waste, establishing stringent monitoring and compliance requirements for waste-to-power plants, requiring minimum reaction temperatures for processing waste materials, establishing rules for dealing with ash and biosolids produced by the energy conversion process, and establishing penalties for noncompliance and failure to meet safety requirements.

Governments and the private sector should invest in R&D to improve the efficiency, safety, and cost effectiveness of waste to energy plants. The efficiency of waste-to-energy plants needs to be improved to ensure that these plants operate as safely and efficiently as possible to ensure that they can produce long-term, reliable energy. This includes investing in modern technologies and improving the performance of existing systems.

In addition, energy storage technologies should be added to waste-to-energy plants because energy storage plays a vital role in stabilizing the energy output of waste-to-energy plants and helping them cope with changes in demand.

Some of these innovative technologies are ready to emerge or in various stages of deployment. For example, the development of steel alloys allows waste-to-power plants to operate at higher thermal efficiencies and generate more power per pound of CO_2 emissions. In addition, several European and other countries have either begun trials of carbon capture systems on their waste-to-power plants or have been making plans for adding such systems.

Raising the public awareness of new features of advanced waste-to-energy plants is also a key to winning their acceptance and support. People tend to be wary of the environmental impact, public health and safety of existing waste-to-power technologies. Educating them on the new safety features and benefits of advanced waste-to-energy plants will play a key role in algae fuel deployment.

4. Algae Fuel Solutions

To make algae a viable renewable energy option, several solutions can be explored. Expanding nonfood-based algae fuel production can help reduce the impact on food production and the environment.

Cultivating microalgae in open raceway ponds or closed photobioreactors offers benefits such as reduced reliance on food crops and job creation. Advancements in technology have focused on improving algae cultivation efficiency and reducing production costs. Designing algae cultivation facilities to capture maximum sunlight, maintaining constant motion in ponds or reactors for optimal growth, and monitoring and controlling nutrient levels, trace elements, and pH of the water are crucial considerations. Adding CO_2 to algae ponds or photobioreactors can enhance algal production. Collaboration and consultation among stakeholders are essential for speeding up technological breakthroughs in algae fuel production.

To address issues related to water and fertilizer use, solutions such as utilizing nutrient-rich wastewater, compost, and organic fertilizers instead of freshwater and employing waste-to-energy technologies such as anaerobic digestion can be explored. Using microbial fertilizers and sustainable alternatives to chemical fertilizers can also contribute to sustainable algae cultivation. Monitoring nutrient levels, implementing stringent wastewater treatment, and adhering to quality standards are important for maintaining algae growth and preventing contamination.

Large-scale experiments are needed to explore productive and efficient algae cultivation options, conversion methods, and fuel performance. Several countries, including the U.S., China, Germany, Australia, Japan, India, and the UK, have recognized the potential of algae fuel and are investing significantly in research and development. For instance, the U.S. Department of Energy is funding research on solar-powered algae biodiesel production and algae-based carbon capture. China is researching algae fuel production, growth methods, and carbon dioxide utilization.

Researchers are also focusing on developing strains with higher lipid content and improving harvesting techniques. New conversion technologies, such as supercritical CO_2 extraction and thermochemical conversion, are being explored to produce higher-quality algae-based fuels at lower costs. Algal biorefining aims to convert algae-based biomass into various products, such as biodiesel, biomethane, and bioethanol. These efforts seek to optimize algae cultivation, enhance fuel efficiency, and promote the commercial production of algae fuels.

5. Solutions for Other Third-Generation Biomasses

The advancement of bioenergy toward its third generation involves a spectrum of strategic solutions that extend beyond algae fuel. These solutions encompass various facets, from innovative seed selection to robust policy frameworks, collectively steering the bioenergy landscape toward sustainability and efficacy.

Selecting and cultivating innovative seed varieties capable of withstanding diverse environmental challenges forms a cornerstone. These varieties should possess attributes such as drought and disease tolerance, resilience to harsh conditions, and compatibility with rotational systems. The development of such versatile seed varieties can bolster large-scale bioenergy production and contribute to consistent yields.

Establishing sustainable large-scale production systems while upholding environmental considerations is paramount. Sustainable agricultural practices, including integrated pest management, soil conservation, water-use efficiency, and erosion prevention, ensure responsible biomass cultivation. These practices safeguard long-term production viability while minimizing ecological impact.

Creating dedicated infrastructure to support perennial biomass fuel production is pivotal. Adequate storage and processing facilities, well-connected transportation networks, and responsive markets collectively facilitate the efficient transition from feedstock to bioenergy. This infrastructure underpins the viability of large-scale biomass production.

The exploration of dedicated energy crops, such as miscanthus and switchgrass, stands as a promising avenue. Research and development efforts focus on optimizing genetics, mixtures, and agronomic practices to enhance biomass yield and quality. This exploration paves the way for resource-efficient and high-yield bioenergy production.

Enabling policy frameworks play a pivotal role in catalyzing private investments and rendering large-scale biomass production economically attractive. Well-designed policies offering incentives and regulatory support create an environment conducive to sustained bioenergy development. These frameworks align economic interests with environmental goals, fostering a robust bioenergy sector.

Elevating public awareness regarding the multifaceted benefits of biomass, spanning environmental preservation, economic growth, and community development, is integral. Public education initiatives demystify bioenergy's significance, nurturing a favorable outlook and bolstering societal support for its proliferation.

Summary

Bioenergy technologies have emerged as a promising renewable energy source with numerous benefits, including energy security, greenhouse gas reduction, waste management, and rural development. Over the years, these technologies have been deployed across various sectors around the world, but they also face challenges that need to be addressed for their future expansion. This chapter explores the benefits of bioenergy technologies, highlights their deployment history, and discusses the challenges and potential solutions for their sustainable growth.

Bioenergy offers several advantages that make it an attractive option for energy production. First, it is a renewable energy source, utilizing organic materials such as biomass and waste that can be continually replenished. Second, bioenergy plays a significant role in reducing greenhouse gas emissions, as the carbon released during combustion is offset by the carbon absorbed during biomass growth. In addition, bioenergy technologies contribute to effective waste management by converting organic waste and agricultural residues into useful energy. Moreover, they enhance energy diversity by reducing reliance on fossil fuels and promoting energy security. Finally, bioenergy projects stimulate rural development by creating job opportunities and supporting local communities.

Bioenergy technologies offer a range of benefits and have a diverse deployment history spanning traditional biomass, first-generation biofuels, waste-to-energy, and advanced biofuels. Traditional biomass has been utilized for centuries for heating, cooking, and fulfilling traditional energy needs, particularly in developing countries. First-generation biofuels, such as ethanol and biodiesel produced from food crops, have been widely implemented, mainly in the transportation sector. Waste-to-energy technologies, including anaerobic digestion, incineration, gasification, and pyrolysis, have been employed to convert organic waste into biofuels, biogas, power, or heat. Moreover, advanced biofuels, such as cellulosic ethanol and algae-based biofuels, have undergone extensive research and development, leading to some commercial-scale deployments.

While bioenergy technologies offer immense potential, they also face certain challenges. The availability and sustainability of feedstock pose a critical concern, necessitating the development of advanced agricultural practices and efficient supply chains. Technological efficiency and scalability are also key challenges that can be addressed through ongoing research, process optimization, and technological innovation. Economic viability is another important aspect, requiring cost reduction measures and financial incentives to make bioenergy projects financially attractive. Environmental impacts, such as land use change, water usage, air pollution, and biodiversity loss, must be mitigated through stringent regulations, sustainability certifications, and advanced emission control technologies. Finally, public acceptance and policy support are vital for the widespread adoption of bioenergy, requiring increased awareness, effective communication, and supportive policy frameworks.

These challenges related to feedstock availability, technological efficiency, economic viability, environmental impacts, and public acceptance need to be addressed for the future expansion of bioenergy. Through continuous research, collaboration, supportive policies, and public engagement, bioenergy technologies can be further developed, ensuring their efficiency, sustainability, and contribution to global renewable energy and sustainable development goals. With concerted efforts, bioenergy will continue to play a pivotal role in the transition toward a sustainable and carbon–neutral future.

Activities

Further Readings

1. Goldemberg, J. & Coelho, T. (2004). Renewable energy—traditional biomass vs. modern biomass. https://doi.org/10.1016/S0301-4215(02)00340-3
2. Butnar, I., et al. (2019). The Role of Bioenergy in Climate Change Mitigation. https://onlinelibrary.wiley.com/doi/full/10.1111/gcbb.12666
3. Ahmad, A. et al. (2011). Microalgae as a sustainable energy source for biodiesel production: A review. https://doi.org/10.1016/j.rser.2010.09.018
4. Vera, I. et al. (2022). Land use for bioenergy: Synergies and trade-offs between sustainable development goals. https://doi.org/10.1016/j.rser.2022.112409
5. Quinn, J. & Davis, R. (2015). The potential and challenges of algae-based biofuels: A review of techno-economic, life cycle, and resource assessment modeling. http://dx.doi.org/10.1016/j.biortech.2014.10.075
6. American Biogas Council (2022). U.S. EPA proposal delivers long-overdue guidance to the biogas industry: ABC. https://www.canadianbiomassmagazine.ca/u-s-epa-proposal-delivers-long-overdue-guidance-to-biogas-industry-abc/

7. Casey, T. (2022). Algae Biofuel Back from Dead, Now with Carbon Capture. https://cleantechnica.com/2022/09/06/algae-biofuel-back-from-dead-now-with-carbon-capture/
8. Yan, M. et al. (2020). Challenges for the Sustainable Development of Waste to Energy in Developing Countries. https://doi.org/10.1177/0734242X20903564
9. Lim, W. et al. (2019). Waste-to-energy: green solutions for emerging markets. https://kpmg.com/xx/en/home/insights/2019/10/waste-to-energy-green-solutions-for-emerging-markets.html
10. Santos, S. et al. (2023). Waste Gasification Technologies: A Brief Overview. https://www.mdpi.com/2813-0391/1/1/11
11. Osman, A. et al. (2023). Materials, fuels, upgrading, economy, and life cycle assessment of the pyrolysis of algal and lignocellulosic biomass: a review. https://doi.org/10.1007/s10311-023-01573-7
12. Dong, J. et al. (2018) Comparison of waste-to-energy technologies of gasification and incineration using life cycle assessment: Case studies in Finland, France and China. https://doi.org/10.1016/j.jclepro.2018.08.139
13. IEA (2020). The outlook for biogas and biomethane to 2040. https://www.iea.org/reports/outlook-for-biogas-and-biomethane-prospects-for-organic-growth/the-outlook-for-biogas-and-biomethane-to-2040

Closed Questions

1. What is bioenergy, and how does it differ from traditional energy sources?
   (a) Bioenergy is energy derived from fossil fuels, while traditional energy sources come from renewable resources
   (b) Bioenergy is energy derived from renewable resources, while traditional energy sources come from fossil fuels
   (c) Bioenergy is energy generated from nuclear reactions, while traditional energy sources come from chemical processes
   (d) Bioenergy is energy generated from wind and solar power, while traditional energy sources come from biomass
   (e) Bioenergy is energy generated from geothermal sources, while traditional energy sources come from hydroelectric power.
2. How is biomass used as a source of energy in traditional bioenergy systems?
   (a) Biomass is directly burned to produce heat and power
   (b) Biomass is converted into liquid biofuels through fermentation
   (c) Biomass is used as a feedstock for nuclear reactions
   (d) Biomass is processed into wind and solar energy
   (e) Biomass is transformed into geothermal energy through underground processes.
3. What are some common examples of traditional biomass used for energy purposes?
   (a) Coal, natural gas, and oil

   (b) Uranium, thorium, and plutonium
   (c) Wood, agricultural residues, and animal manure
   (d) Solar panels, wind turbines, and hydroelectric dams
   (e) Geothermal heat pumps, tidal power, and wave energy converters.
4. What are the advantages of using traditional biomass for energy production?
   (a) Abundant resources and carbon–neutral emissions
   (b) Low-cost energy, reduced environmental impact
   (c) Energy security, reduced reliance on fossil fuels
   (d) Technological advancements and energy efficiency
   (e) Reduced greenhouse gas emissions, sustainable land management.
5. What are the disadvantages of using traditional biomass for energy production?
   (a) Prohibitive cost, limited availability
   (b) High processing need, water use
   (c) Limited scalability, high maintenance
   (d) Deforestation and air pollution
   (e) Inconsistent energy output, high water consumption.
6. How does modern biomass differ from traditional biomass in terms of energy generation?
   (a) Modern biomass utilizes advanced conversion technologies, while traditional biomass relies on direct combustion
   (b) Modern biomass has lower carbon emissions, while traditional biomass produces significant greenhouse gases
   (c) Modern biomass is derived from renewable sources, while traditional biomass is derived from fossil fuels
   (d) Modern biomass requires less land and water resources than traditional biomass
   (e) Modern biomass is used exclusively for electricity generation, while traditional biomass is used for heat and power.
7. Which of the following is not one of the advanced bioenergy technologies and processes?
   (a) Gasification
   (b) Anaerobic digestion
   (c) Combustion
   (d) Pyrolysis
   (e) Torrefaction.
8. Which of the following processes of algae as a source of biofuel is best aligned with the advantages of algae fuel?
   (a) Algae is converted into biodiesel through transesterification
   (b) Algae undergo gasification to produce syngas
   (c) Algae is fermented to produce ethanol
   (d) Algae is used to produce biogas through anaerobic digestion
   (e) Algae is processed to extract oils for biofuel production.

9. Which of the following is not a potential environmental benefit of using bioenergy derived from biomass?
   (a) Reduction of greenhouse gas emissions
   (b) Decreased reliance on fossil fuels
   (c) Preservation of biodiversity
   (d) Promotion of sustainable land management practices
   (e) Mitigation of air pollution.
10. In which of the following aspects can bioenergy not contribute to reducing greenhouse gas emissions and mitigating climate change?
   (a) Replacing fossil fuel combustion in electricity generation
   (b) Replacing fossil fuel use in transportation
   (c) Increasing deforestation and land degradation
   (d) Replacing fossil fuel use in heating and cooling systems
   (e) Promoting carbon capture and storage technologies.
11. Which of the following is not one of the challenges and limitations associated with the widespread adoption of bioenergy?
   (a) Competing land use for food production
   (b) Unstable sustainable feedstock supply
   (c) Lack of technological and economic feasibility
   (d) Environmental and social impacts
   (e) Policy and regulatory support.
12. How can bioenergy and biomass play a role in achieving sustainable and renewable energy goals globally?
   (a) By increasing greenhouse gas emissions and environmental degradation
   (b) By supporting energy diversification and reducing reliance on fossil fuels
   (c) By depleting natural resources and contributing to deforestation
   (d) By exacerbating climate change and global warming
   (e) By hindering a transition to cleaner and more efficient energy systems.

Open Questions

Biomass:

1. What are the most sustainable and efficient methods for biomass production and utilization?
2. How can we optimize the conversion processes of different types of biomass into energy or valuable products?
3. What is the environmental impact of large-scale biomass harvesting and utilization?
4. How can we improve the logistical challenges associated with biomass collection, transportation, and storage?

Bioenergy:

1. What are the economic and environmental benefits of bioenergy compared to traditional fossil fuels?
2. How can we enhance the efficiency and cost-effectiveness of bioenergy conversion technologies?
3. What are the potential social and cultural implications of widespread bioenergy adoption?
4. How can bioenergy contribute to energy security and the decentralization of energy production?

Municipal solid waste (MSW) to power:

1. What are the most effective strategies for MSW management and waste-to-energy conversion?
2. How can we optimize the sorting and recycling processes to maximize resource recovery from MSW?
3. What are the environmental and health impacts of different MSW disposal methods, such as landfilling or incineration?
4. How can public perception and acceptance of MSW-based bioenergy be improved?
5. Food-based first-generation bioenergy:
6. What is the trade-off between food production and bioenergy production using food crops?
7. How can we ensure sustainable sourcing of feedstocks for food-based bioenergy to avoid competition with the food supply?
8. What are the social and economic consequences of diverting food crops for bioenergy production?

Waste-based bioenergy:

1. How can we optimize the conversion of different waste streams (e.g., agricultural residues, industrial waste, municipal solid waste) into bioenergy?
2. What are the best practices for waste management and utilization to maximize bioenergy production and minimize environmental impact?
3. What are the potential challenges and opportunities of waste-based bioenergy in different regions and industries?

Algae fuels:

1. How can we improve algae cultivation methods and productivity for biofuel production?
2. What are the best extraction and refining techniques to obtain high-quality biofuels from algae?

3. What is the overall life cycle impact of algae fuels, including their carbon footprint and environmental sustainability?
4. How can algae fuel technologies be scaled up for commercial production?

Geothermal Power and Heating

6

Introduction

Science and Technology

1. Geothermal Energy

Geothermal energy is a renewable heat source inside the Earth that can be used to generate power in geothermal power plants, heat or cool buildings, or conduct many other economic activities.

Geothermal energy comes from the slow decay of radioactive particles in the Earth's core. The temperature of the inner solid core of nickel and iron (approximately 2440 km (1516 miles) in diameter) is approximately 6000 °C (10,832 °F), which is even higher than that of the surface of the sun, i.e., 5500 °C (9932 °F). This intense heat causes material in the molten outer core (magma) of nickel and iron (approximately 2440 km (1516 miles) thick) and the mantle of semisolid silicate rock (approximately 2900 km (1802 miles) thick) to move around (convection). The convective temperature decreases from 4000 °C (7232 °F) for the outer core to 200 °C (392 °F) for the mantle.

The Earth's outermost layer (approximately 35–70 km or 22–43 miles thick) of mostly rock granite is called the crust, which is dissected into pieces called tectonic plates. The fault lines of the plates are easy passageways for magma to advance to the Earth's surface, which not only causes the eruption of lava through the Ring of Fire's volcanoes but also heats underground Earth and water and provides abundant geothermal resources in those regions. In addition, the crust itself maintains a decreasing conductive temperature from 200 °C (392 °F) at the edge of the much hotter mantle to the Earth's surface at a "geothermal gradient" rate of 25–30 °C

P. Yang, *Renewable Energy*, https://doi.org/10.1007/978-3-031-49125-2_6

(77–86 °F)/km. This gradual temperature change allows a relatively stable underground soil temperature of 21–24 °C (70–75 °F) at a depth of 4 m without direct exposure to magma, and the air temperature changes above the Earth throughout the seasons.

In other words, except for the top one meter (3.3 ft) of land heated by sunlight, geothermal energy is a relatively self-sustained renewable energy source mainly affected by the Earth's core. Therefore, geothermal energy is different from other renewable energy sources—such as solar, wind, and hydro—that all result from solar radiation and energy. Underground heat comes from the outward extension of the extreme heat caused by the decay of radioactive particles in the Earth's interior core. However, similar to solar thermal energy, geothermal energy as a thermal energy source (steam or hot water) can be used either as moving energy to drive the turbine for power generation (geothermal power generation) or as a thermal energy source to heat, cool, cook, dry, etc. (geothermal heat direct use).

2. Geothermal Power Generation

Geothermal Power Generation Technologies. Like most other renewable or conventional power generation technologies, such as hydropower, wind power, concentrated solar power, fossil fuel-fired power, and nuclear power, turbines and generators are used for geothermal power generation. Figure 6.1 shows the differences between the three types of geothermal power generation technologies in science and technology, stage, application, and efficiency.

However, geothermal power generation requires wells to harness subsurface heat and there are different underground geological conditions, three types of conventional geothermal power generation technologies—dry steam, flash steam, and binary cycle geothermal systems are used.

The existing types of geothermal power plants use progressively innovative technologies. The first two types—dry steam and flash steam—use natural steam from the geothermal reservoir, but the flash steam geothermal power plant differs from the dry steam plant in two aspects. First, it also uses high-temperature water in addition to steam. Second, the geothermal water is flashed to steam, and the remaining water droplets are separated from steam before steam drives the turbine and generator to produce power.

The third type—a binary cycle geothermal power plant—differs from the first two types in its design of two separate closed loops, which allows geothermal resources in the first closed loop to transfer heat to a low boiling point working fluid in the second closed loop, causing the latter to turn to steam for power generation and then return in the first closed loop to the reservoir without impacting the environment during power generation.

Unlike variable solar and wind power generation, geothermal power generation is consistent and always operational and is therefore considered steady-state baseload power. Geothermal power plants generate power 24 h a day, 7 days a week, and their power output is highly predictable and stable. In addition, geothermal power generation can be easily turned on and off. The consistency

	Dry Steam	Flash Steam	Binary Cycle
Science & Technology	Steam from a geothermal reservoir turns the turbine / generator to generate power	High-pressure fluid from the reservoir pumped into a tank at a much lower pressure; resulting flash steam turning the turbine / generator to generate power (single flash); remaining fluid usable for double or triple flash	Hot water or steam running in a closed loop in parallel to another closed loop of working fluid with a much lower boiling point; heat moving from the geothermal loop to the working fluid; resulting steam from working fluid driving the turbine / generator to generate power; spent water returning to the ground
Open/Closed Loop	Open	Open	Closed
Location Restriction	Yes	Yes	Yes
Geothermal Power Technology	An early type, first developed at Lardarello, Italy in 1904	An early type, first developed at Lardarello, Italy in 1904	A recent type, first developed in Ischia, Italy, during 1940–1943
Application	Still used today, at largest power plants at approximately 205 °C (401 °F)	Most used today	Increasingly, but not yet widely used because of high expenses
Requirement	Geothermal sites with steam	Water and steam at temperatures of 174–315 °C (345–600 °F)	A dual closed-loop heat transfer system is used, one for geothermal fluid and the other for working fluid with a low boiling point (isobutane, freon-12, ammonia, or propane)
Efficiency	Low efficiency because it only relies on naturally available resources	Higher efficiency as it uses hot water to generate steam to generate power	Highest efficiency as it generates power using a fluid at lower temperatures of 150–205 °C (300–400 °F)
Sustainability, Environmental Impacts	Issues with sustainability, reliable operation, and environmental friendliness	Issues with sustainability, reliable operation, and environmental friendliness	Allows sustainability, high-reliability operation, and environmental friendliness

Fig. 6.1 Types of traditional geothermal power technologies

and flexibility of geothermal power generation allow it to qualify to serve as an intentionally variable power source or a load-following power source, i.e., both a base-load power plant generating less power during off-peak and a peaking power plant generating more power in quick response to peak demands.

3. Geothermal Heat Direct Use

Geothermal heat is the natural replenishment of hydrothermal resources, such as steam or hot water, within the Earth that can be harnessed through various technologies. In addition to geothermal power generation, geothermal heat can also be directly used in various applications, depending on the resources and technologies chosen, such as heating and cooling buildings, industrial drying processes, and greenhouses, through geothermal heat pumps. Geothermal heat can be used to replace old or develop new district heating and cooling systems in residential areas, industrial complexes, hotels, office buildings, hospitals, and other large energy end applications.

Benefits

1. Renewable Energy Source

Similar to solar, wind, and hydro energy, geothermal energy is one of the renewable energy sources widely accepted as a clean and sustainable alternative to fossil fuels. The main differences between geothermal energy and fossil fuels include sustainable availability, significant CO_2 reduction, environmental protection, and reduced operating cost. Geothermal energy is sustained by the massive heat from the Earth's interior. The direct use of geothermal heat is very sustainable in terms of both the tiny amount of its use compared to the enormous availability of heat sources in the Earth's core and the sharp contrast between its minor environmental impact and that of the use of fossil fuels. The latter is formed from the remains of dead animals and plants over millions of years. They are mined and drilled from the ground and will run out in a much shorter period in Earth's history.

2. Environment and Health Benefits

Although geothermal power and heat generation have a minor environmental impact, this impact is insignificant in comparison to the use of fossil fuels. For the same amount of power generated, the use of geothermal energy can displace most greenhouse gas emissions from fossil fuel-fired power generation—95.3% of greenhouse gases emitted using coal, 94.7% using oil, and 92.2% using natural gas. Therefore, using geothermal energy for power generation plays a significant role in our climate action to meet the global Net Zero target.

Since geothermal power plants do not burn fossil fuels to generate power, there are no emissions of carbon dioxide or other air pollutants from burning fuel. Although well drilling, manufacturing, installing and operating geothermal plants release some chemicals, these emissions of geothermal power and heat generation and direct use are minor and can in most cases be controlled and minimized through appropriate mitigation measures. Indeed, these emissions are insignificant compared with those from power generation using fossil fuels. This means that geothermal power generation can offset 97% of acid rain-causing sulfur compounds and approximately 99% of carbon dioxide from fossil fuel power plants of comparable size. Geothermal power plants also use technology such as scrubbers to remove the hydrogen sulfide naturally found in geothermal reservoirs. Most geothermal power plants inject condensed geothermal steam and water back into the Earth, which recycles the geothermal resources and reduces emissions from the geothermal power process.

3. Flexible Power

Geothermal power generation is different from other renewable energy sources, such as solar and wind, which can vary depending on weather and environmental conditions. Unlike these sources, geothermal power can generate a constant and reliable source of power, making it a "baseload" power. This means that geothermal plants can operate consistently, providing a continuous and stable supply of electricity to the grid, such as hydropower generated by a dam and reservoir.

Additionally, geothermal power plants are highly flexible and can start and stop quickly, making them well suited to meet peak demand. This makes geothermal power not only a baseload power but also a "peak load power" source. Peak load power is required during high electricity demand periods when additional electricity is needed to supplement the baseload power supply.

By increasing the share of geothermal power in the energy mix, we can provide a robust renewable baseload power source that can displace fossil fuels and add renewable peak load power to balance and stabilize the power supply.

4. High Capacity Factor

Another benefit of geothermal power generation is its high capacity factor. The capacity factor measures the actual power generation output of a power plant as a percentage of its maximum potential output (called power generation capacity). It is an important performance indicator for geothermal power plants. The capacity factor of geothermal power plants ranges from 70 to 95%, which is much higher than most other renewable energy technologies and comparable to conventional fossil fuel power plants.

The higher capacity factor of geothermal power generation compared to other renewable energy sources is possible for several reasons. The primary reason is geothermal power generation's feature as flexible power described above. Earth's internal heat provides a constant source of energy, and geothermal power plants can use it to generate power continuously, 24 h a day, 7 days a week, and power generation is not affected by weather conditions or daily fluctuations.

In addition, geothermal power plants have a high availability factor, meaning they are often able to operate at their full capacity. This is because they do not rely on external factors such as fuel supply or weather conditions.

Furthermore, geothermal power plants have a longer lifespan than other renewable energy sources. Once a geothermal reservoir is developed, it can produce power for 20–30 years or more. This allows for a stable and reliable source of energy that can contribute to meeting the world's energy demands.

Deployment

A. Geothermal Technology Growth Dynamics

As one of the oldest green energy alternatives for fossil fuels, geothermal power plays a significant role in our energy transition and carbon reduction. Humans have been using heat under the Earth's surface to perform practical and spiritual activities for at least ten thousand years, such as using hot springs for hot bathing, healing, heating in winter, and food cooking, as well as religious and cultural activities. Geothermal heating, using water from hot springs, for example, has been used for bathing since Paleolithic times and for space heating since ancient Roman times. More recently, the use of geothermal energy to generate electric power and various heating applications has gained importance as an important way to achieve Net Zero goals. It is estimated that the Earth's geothermal resources are theoretically more than adequate to supply humanity's energy needs, although only a tiny fraction is currently being profitably exploited, often in areas near tectonic plate fault lines.

1. Geothermal Power Generation

The first effort to harness geothermal energy for industrial use came in 1818 in the Tuscan region of Italy. French engineer François Jacques de Larderel pioneered a new way to extract boric acid from hot springs. While others had developed the means to extract the acid using fire to evaporate the water, he was the first to harness the region's substantial geothermal energy to drive the process. The town that grew up around the industrial production of boric acid, Larderello, was also home to the first successful effort to produce electricity with geothermal energy. Eighty-six years later, Italian scientist Piero Ginori Conti succeeded in using geothermal energy to generate power for several light bulbs, which led to much larger experiments and the operation of the world's first commercial geothermal power plant, Larderello, Italy, in 1914.

After the Second World War, the U.S. became a major producer of geothermal power. The world's largest geothermal power plant complex is The Geysers, located in the Mayacamas Mountains north of San Francisco. Opened in 1960, the site became the world's largest complex and now includes thirteen power plants operated by dry steam from over 350 wells.

Since then, Indonesia, the Philippines, Türkiye, New Zealand, Mexico, Kenya, Iceland, and Japan have also exploited geothermal resources for power generation

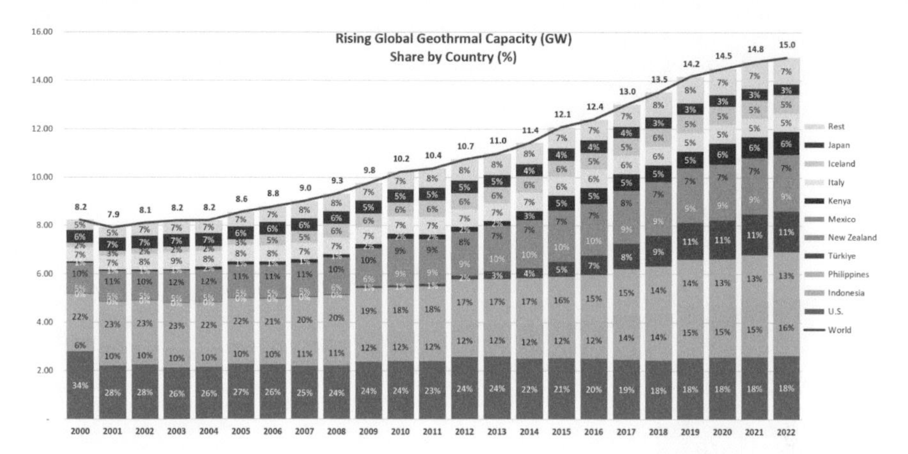

Fig. 6.2 Global geothermal power capacity in 2022. *Data Sources* Our World in Data, IRENA: 2023

with continued growth. According to statistical data from the International Renewable Energy Agency (IRENA), the U.S. remains the world leader in geothermal installed capacity (2.7 GW), followed by Indonesia (2.3 GW) and the Philippines (1.9 GW) in 2022. Combined, the top three geothermal power countries accounted for almost a half of the world's total geothermal power generation capacity, and the top 10 countries were responsible for 93% of the global total geothermal power capacity (see Fig. 6.2).

Although geothermal power has a longer history than wind power and solar power, its growth has been slow, and it remains a much smaller renewable energy industry than wind power and solar power. In 2022, geothermal power represented 0.4% of power generation in the largest geothermal power generating country, the U.S.

Approximately 500 geothermal power plants are generating power in the world, with a total installed power generation capacity of more than 15.7 gigawatts (GW) in 2022. Geothermal power plants are smaller than hydropower plants in size. As of 2022, 76% of all geothermal power plants across the world have a power generating capacity smaller than 42 MW, 20% of them have a power generating capacity between 42 and 112 MW, and only 4% of them have a capacity larger than 112 MW (see Fig. 6.3).

From the technological perspective, dry steam power stations are older and larger than other types of geothermal power stations—dry steam is used by 13% of geothermal power stations, but these power plants account for 23% of the global installed geothermal power capacity.

The majority (44%) of the geothermal power plants use flash steam technology, and they account for an even larger share of the total installed power generation capacity (56%), indicating that these flash steam power plants are also larger than average-size geothermal power plants.

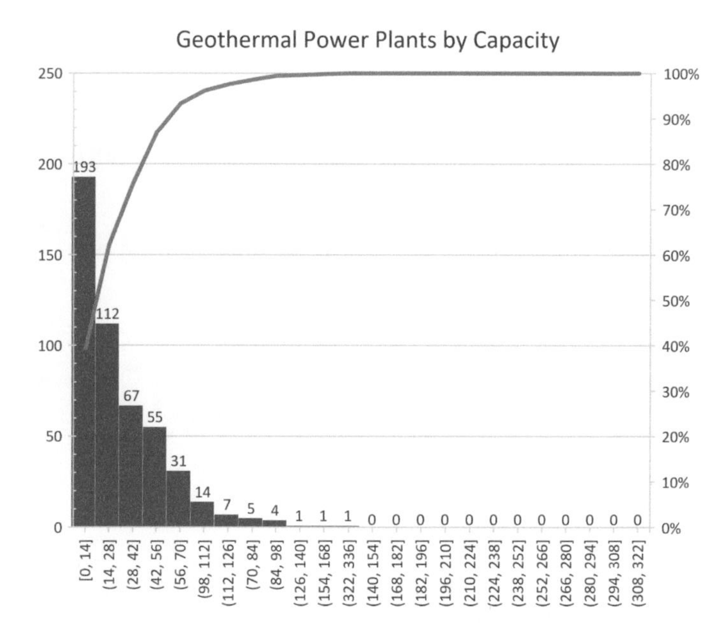

Fig. 6.3 Global geothermal plants by capacity. *Data Source* Wikipedia

A substantial number (38%) of geothermal power plants use binary cycle technology, but they account for a much smaller share of the total installed power generation capacity (18%), indicating that binary cycle power plants are generally smaller than average-size geothermal power plants (see Fig. 6.4). In addition, there are many experimental projects of enhanced or engineered geothermal systems (EGS). However, there is no EGS in commercial operation as of 2023.

Of the top 10 geothermal power stations in the world, two in the U.S. account for 13% of the world capacity using dry steam technology, i.e., the oldest geothermal power technology in the world, and one in the U.S. accounts for 2% using the flash steam technology, i.e., the currently most prevailing geothermal power technology.

The stations in Mexico, Italy, and Kenya account each for 5% of the global geothermal power capacity using dry steam technology, and those in New Zealand, Indonesia, Philippines, and Iceland have each 2% of the global geothermal power capacity using flash steam technology. In other words, most (28% of total capacity) of the top 10 geothermal power stations are using the oldest geothermal power generating technology and the stations using the most prevailing technology only account for 10% of the global capacity (Fig. 6.5).

There are approximately 40 manufacturers of geothermal power turbines in the world. The top five manufacturers supplied geothermal power turbines for 73% of all geothermal power plants with a combined geothermal power generation capacity of 12 GW, which accounts for 86% of the global total installed geothermal power generation capacity.

Technology	Power Stations Using It	Share in Total Power Stations	Capacity Using It	Share of Total Capacity
Dry Steam	65	13%	3244.4	23%
Flash Steam	218	44%	7743.9	56%
Single Flash	158	32%	4988.1	36%
Double Flash	58	12%	2566.8	18%
Triple Flash	2	0%	189.0	1%
Binary Cycle	187	38%	2490.3	18%
Flash & Binary Cycle	1	0%	330.0	2%
Back Pressure	19	4%	115.4	1%
Co-Production	1	0%	0.1	0%
Total	491		13924.2	

Fig. 6.4 Global geothermal power plants by technology. *Data Source* Wikipedia

Station	Country	Capacity (MW)	Plants	Type	Share in Total Capacity
The Geysers	U.S.	1590	13	Dry steam	10%
Cerro Prieto	Mexico	820	5	Dry steam	5%
Larderello	Italy	769	34	Dry steam	5%
Olkaria	Kenya	727	5	Dry steam	5%
Imperial Valley	U.S.	403	11	Dry steam	3%
Wairakei	New Zealand	352	4	Flash steam	2%
Sarulla	Indonesia	330	3	Flash steam	2%
Tiwi	Philippines	330	3	Flash steam	2%
Hellisheiði	Iceland	303	7	Flash steam	2%
Coso	U.S.	270	4	Flash steam	2%

Fig. 6.5 Top 10 geothermal power stations. *Data Source* Wikipedia

Figure 6.6 shows information about these manufacturers' turbine supplies, including the number of geothermal power plants they supplied, the capacities of their supplied turbines, and these turbines' shares in the global geothermal power capacity, as well as the geothermal power generation technologies (dry steam, flash steam, binary cycle, etc.) on which these geothermal turbine manufacturers focused.

It shows that, Toshiba, a mixed technology turbine manufacturer, has only supplied to a small number of geothermal power plants (43), accounting for 9% of total geothermal power stations, but it is the biggest geothermal turbine supplier in terms of geothermal power generation capacity (3163 GW), i.e., 23% of the

Turbine Manufacturer	Number of Stations Supplied	Share in All Geo Power Stations	Capacity Supplied (MW)	Share in All Capacity	Technology Provided
Toshiba	43	9%	3,163	23%	Mixed
Fuji	66	13%	2,783	20%	Flash Steam
Mitsubishi	81	16%	2,853	20%	Flash Steam
Ormat	127	26%	2,025	15%	Binary Cycle
Ansaldo/Tosi	43	9%	1,173	8%	Mixed
Total	360	73%	11,997	86%	

Fig. 6.6 Top 10 geothermal power turbine manufacturers. *Data Source* Wikipedia

total geothermal power generation capacity in the world. In contract, Ormat, a manufacturer of the more advanced binary-cycle geothermal technology, is the top geothermal turbine supplier in terms of total number of geothermal power stations it has supplied to (127), which is almost three times of Toshiba's, however, the total geothermal power capacity it has supplied is only two thirds of Toshiba's. This indicates that Toshiba supplies mainly technologically older and larger geothermal power stations while Ormat serves more advanced yet smaller binary-cycle geothermal power stations.

2. Geothermal Direct Use

The history of the direct use of geothermal energy is rich and varied. Hot springs have been used for bathing at least since Paleolithic times. Native Americans have used hot springs in Yellowstone in the U.S. for religious and medicinal purposes for more than 10,000 years. Romans used hot springs to feed public baths and underfloor heating in Bath, England, in the first century CE. The world's oldest geothermal district heating system in Chaudes-Aigues, France, has been operating since the fifteenth century. The earliest industrial exploitation began in 1827 with the use of geyser steam to extract boric acid from volcanic mud in Larderello, Italy. The first U.S. district heating system in Boise, Idaho, directly used geothermal heat in 1892, which was copied in Klamath Falls, Oregon, in 1900.

In the twentieth century, geothermal direct use began to develop rapidly as a viable source of energy in Europe and the U.S. One notable example is the development of geothermal district heating in Reykjavik, Iceland, which began in the 1930s. The city sits on top of a geothermal reservoir, and the hot water and steam from the underground reservoir are used to heat buildings and provide hot water to residents. This district heating system has been expanded and refined over the years and now provides heat and hot water to virtually all of Reykjavik's population.

Another example is the use of geothermal energy for greenhouse heating in the Netherlands, which began in the 1970s. The Netherlands has a large horticultural industry, and geothermal energy has proven to be a cost-effective and sustainable way to provide heat for greenhouse cultivation. This industry has since grown

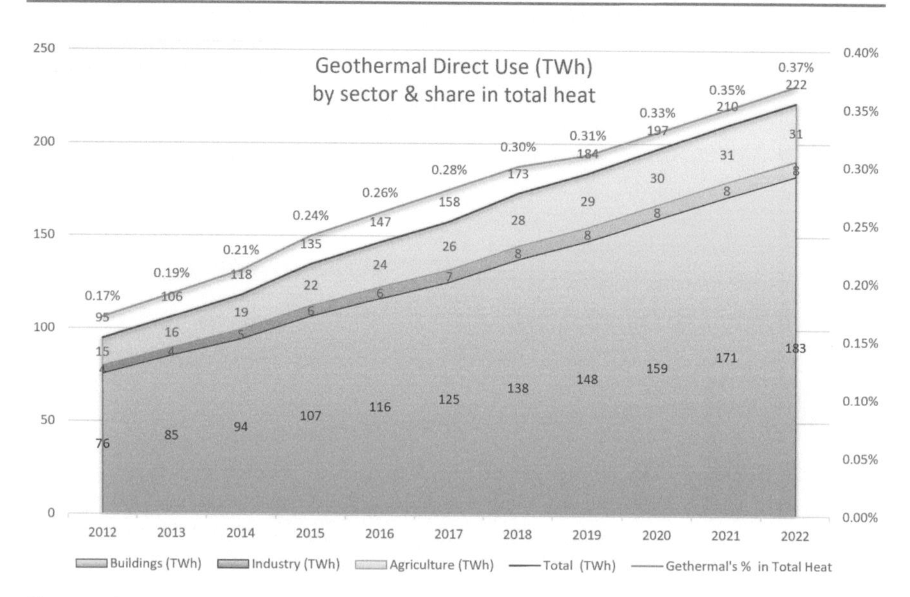

Fig. 6.7 Global geothermal direct use. *Data Source* IEA, CC BY 4.0

significantly, and the Netherlands is now a world leader in the use of geothermal energy for greenhouse heating.

Technological advances have dramatically reduced costs and thereby expanded the range and size of the direct use of viable geothermal resources. According to the data of the International Energy Agency (IEA), the direct use of geothermal energy increased by 134% from 95 TWh in 2012 to 222 TWh in 2022, accounting for an average annual growth rate of 13% (Fig. 6.7).

In the U.S., the direct use of geothermal energy began to take off in the 1970s with the development of geothermal district heating systems in places such as Klamath Falls, Oregon, and Boise, Idaho. These systems provided heat to public buildings, schools, and other facilities in their respective communities.

In recent years, countries such as China, Türkiye, and Kenya have also seen rapid growth in the direct use of geothermal energy. In China, geothermal district heating systems have been installed in cities such as Beijing, Tianjin, and Xian. In Türkiye, geothermal energy is used for both district heating and greenhouse cultivation, and the country has set a goal to increase its installed geothermal capacity to 1 GW by 2023. In Kenya, geothermal energy is used to generate electricity and provide heat to rural communities. The country has seen significant growth in its geothermal industry in recent years, with its installed geothermal capacity increasing from 198 MW in 2015 to 692 MW in 2020.

Overall, the rapid development of geothermal direct use has been driven by a combination of factors, including the availability of geothermal resources, advances in technology, and a growing recognition of the need for sustainable energy sources. The higher roundtrip efficiency of geothermal direct use compared

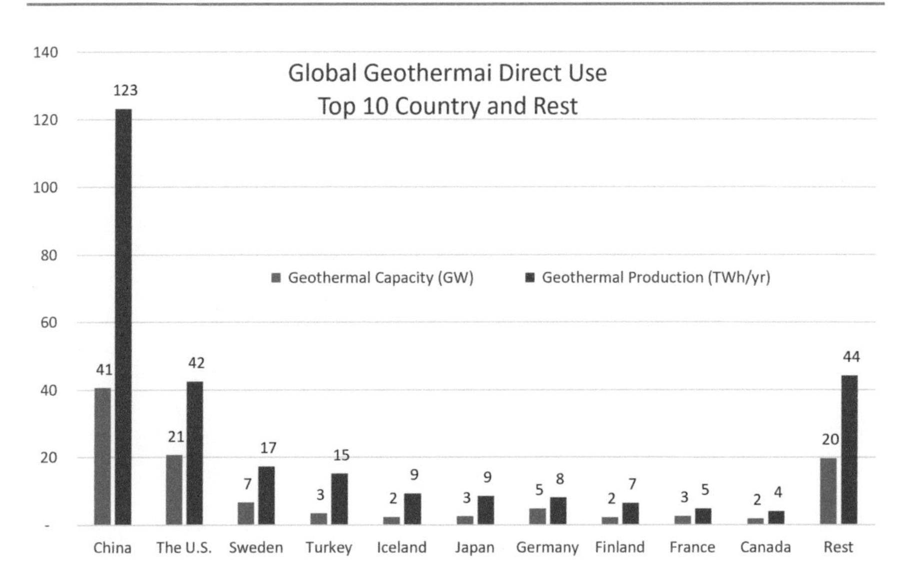

Fig. 6.8 Geothermal direct use by country: top 10 and the rest. *Data Source* Lund, J. & Toth, A., 2021 Geothermics

to geothermal power generation for space and hot water heating applications has also contributed to its increasing popularity.

China is the world leader in the direct use of geothermal heat, with a geothermal heat capacity of 41 GW and an annual geothermal production of 123 TWh. The U.S. is the world's second largest country in the direct use of geothermal heat, with 21 GW of geothermal heat capacity and 42 TWh of annual geothermal production. Sweden is the third largest country in the direct use of geothermal heat, with 7 GW of geothermal heat capacity and 15 TWh of annual geothermal production.

Germany and Japan have geothermal heat capacities of 5 and 3 GW, respectively. Türkiye, Iceland, France, Finland, and Canada all have geothermal heat capacities of 3 GW or less. While Iceland and Japan produce approximately 9 TWh annually, Germany, Finland, France, and Canada all produce less than 10 TWh per year. The combined geothermal capacity of the rest of the world is 20 GW, while their combined geothermal production is 44 TWh per year (see Fig. 6.8).

Challenges

There are many challenges facing geothermal energy technologies, especially geothermal power generation technologies, causing the deployment of geothermal power generation at a slow pace over an extended period. According to the statistical data from IRENA, the average annual growth rate of the installed geothermal power generation capacity was less than 5% in the last 10 years and 4% in the last 20 years (Fig. 6.9).

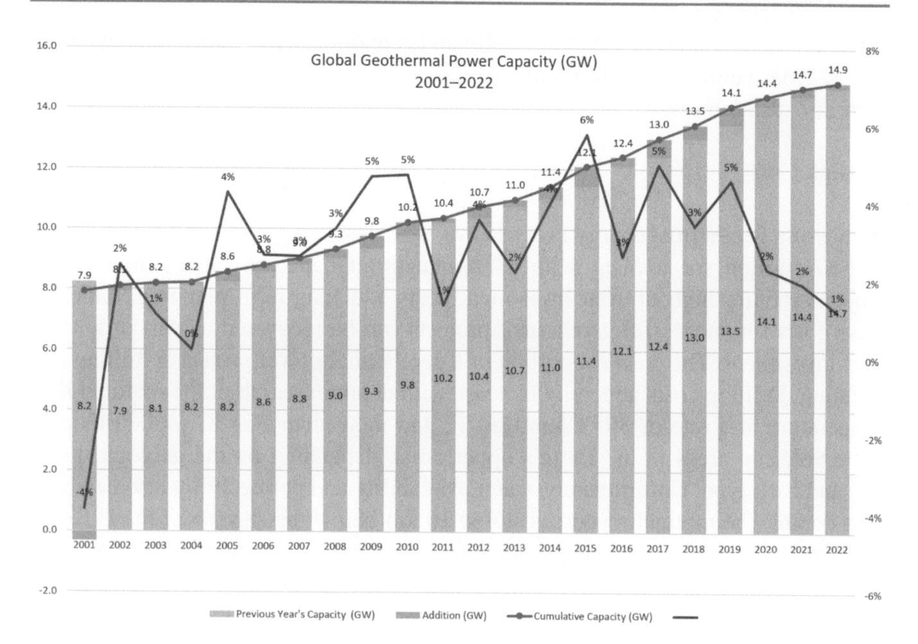

Fig. 6.9 Global installed geothermal capacity (GW). *Data Source* IRENA, 2023

A. Technological Challenges

1. Location Restrictions

The geothermal sector has traditionally been dominated by regions where high-enthalpy hydrothermal reservoirs are available at depths within 3 km (2 miles). Looking at the active geothermal wells for power generation, most have reservoir temperatures of more than 200 °C (400 °F). Given these temperature requirements, the geothermal location factor has proven to be a bottleneck for the scalability of geothermal energy because the number of countries that have high-temperature reservoirs located at those depths is extremely limited.

The technological and economic requirements of traditional geothermal technologies constrain the locations of geothermal power and heat production. Geothermal power and heat plants need to be built in places where hydrothermal resources such as steam and hot water are available and economically accessible. Over 90% of the global geothermal energy generation potential is localized in only 10 countries. This means that most inhabited areas on Earth are either not endowed with all three required geothermal resources or are unable to access and exploit these resources economically.

2. Technological Requirements

Traditional geothermal technologies—dry steam, flash steam, and binary cycles—all require heat, water, and high permeability. These three requirements strictly constrain the potential of technologically and economically extractable geothermal energy from below the Earth's surface. Accessible abundant geothermal resources with dry steam are rare.

3. High Costs

Harnessing geothermal energy is expensive, with costs ranging from approximately \$2—\$7 million for a plant with a one-megawatt capacity. Geothermal power projects require high upfront investment costs and long-term planning, although these costs for successful geothermal projects can be recovered as part of a long-term investment. The costliest portion of a geothermal power project is drilling and planning. Drilling costs generally increase with depth in search of the required heat temperature at a rate of 30 °C/km (86 °F) and more than 200 °C (400 °F)/km in some volcanic areas in the Earth's upper crust, depending on the geological setting and rock type. Unfortunately, water availability and permeability also decrease with depth. The thermal resources stored 10 km below the surface of the U.S. are estimated to exceed 100 times the annual energy demand of the country. However, the current drilling technology allows wells to be drilled routinely to 5 km, and the current practical limit of drilling depth is approximately 10 km (16 miles).

Although it is technically possible to build geothermal power plants anywhere on Earth using enhanced or engineered geothermal systems (EGS), these power generation technologies are currently not as economically feasible as technologies using other renewable energy sources. The geothermal temperatures of geothermal fields are inversely related to the costs of geothermal energy in those areas. In areas where geothermal reservoirs naturally reach 250 °C (482 °F), geothermal energy costs only \$1000/kW. In areas where the geothermal resources are only approximately 100 °C (212 °F), the geothermal energy cost rises sixfold to \$6000/kW. The sharp contrast in the rising cost of low-temperature geothermal power generation compared to that of high-temperature power generation is caused by two factors. First, low-temperature geothermal technology needs to engineer and construct an artificial geothermal reservoir. Second, low-temperature geothermal technology lacks naturally moving energy and therefore must work at a slower rate of energy generation.

In contrast to solar PV and wind power, there was no significant cost reduction in the last ten-year period of 2012–2021 for geothermal projects. Historical data on the global levelized cost of energy for geothermal power indicate that geothermal power remained stable at \$70 per MWh (see Fig. 6.10).

In addition, the future costs of geothermal power are also much higher than those for solar PV power generation with tracking or storage and wind power generation. The capital cost for a 50 MW geothermal power plant is 1.5 times that for a 200 MW wind farm, more than 1.7 times that for a 150 MW solar farm with storage, and more than 2.2 times that for a 150 MW solar farm with tracking. In addition, geothermal power generation also has significantly higher operating

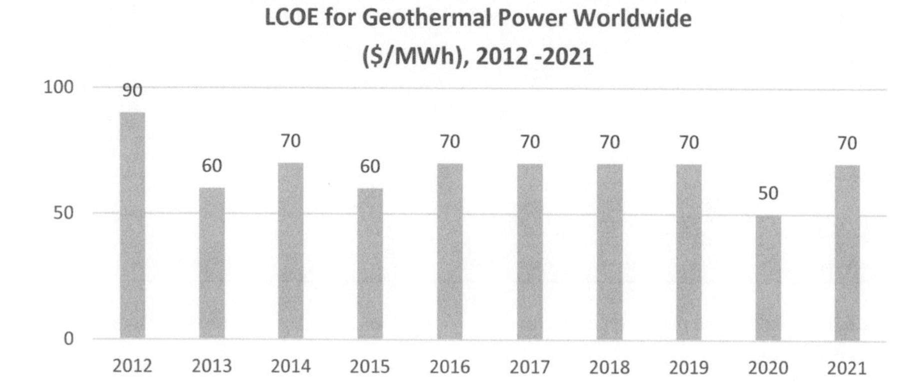

Fig. 6.10 Historical costs of geothermal power. *Source* Statistica 2023, IRENA 2022, Survey

Technology	Capital cost (2020 $/kW)	Size (MW)	Variable operating & maintenance (2020 $/MWh)	Fixed operating & maintenance (2020 $/kW-y)	Heat rate (Btu/kWh)	Year first available
Solar PV w tracking	1,248	150	0	15.33	NA	2022
Solar PV w storage	1,612	150	0	32.33	NA	2022
Wind	1,846	200	0	26.47	NA	2023
Geothermal	2,772	50	1.17	137.5	8,946	2024

Fig. 6.11 Costs of future power generation technologies. *Source* NREL

and maintenance costs, 3.2 times higher than solar PV with storage, 4.2 times higher than wind power, and eight times higher than solar PV with tracking (see Fig. 6.11).

4. Ecosystem Impacts

Because the gas emissions from geothermal plants are far lower than those associated with fossil fuel-fired power and heat plants, geothermal energy is generally considered clean energy. However, it has gas emissions and other ecosystem impacts in both the well-drilling and operational phases and, therefore, is considered not as clean as solar and wind energy.

Geothermal power generation projects require energy-intensive drilling to source geothermal resources, and the large-scale development of geothermal energy projects will cause emissions and the ecosystem because of its use of fossil fuels. The operations of geothermal power and heat generation do not burn fossil fuels, but they also release—long with hydrothermal fluids—low amounts of carbon dioxide, hydrogen sulfide, and ammonia, as well as other potential pollutants such as mercury, boron, and arsenic, which are stored under the Earth's surface. While these gases are also released into the atmosphere naturally, the emission rates increase near geothermal power and heat plants. This issue is especially obvious for conventional open-loop geothermal power and heat plants.

	Flash	Combined Heat & Power (CHP)	HeatORC	EGS
Geothermal mover	Steam	Water/Steam	Working fluid	Water
Technology	Self-flowing	Self-flowing	Downhole pumps	Downhole pumps
Power/heat generation unit	Flash steam plant	Double flash, combined heat, and power plant	Binary, heat exchanger	Heat exchanger
Cooling system	Wet cooling tower	Wet cooling tower	Air cooling tower	None
Direct emissions	7% in mass of the flow rate of the geothermal fluid average gas fraction and composed of 92% CO_2	0.12–0.23% CO_2 of the flow rate of the geothermal fluid 0.00–0.21% CH4 of the flow rate of the geothermal fluid	None	0.1–2% of the flow rate of the geothermal fluid direct emissions with very small amounts of CO_2 (0–1%) and CH4 (0-0.01%)
Gas control system	NCG abatement system	None	None	None
Final energy use	Power + Industrial heat	Power + Heat	Heat (+power for self-consumption)	Industrial heat

Fig. 6.12 Geothermal plants: types, technologies, emissions, and uses. *Data Source* Douziech, M. et al. Models, 2020

In general, the emission rates of geothermal power and heat generation in the operational phase differ significantly depending on which type of power and heat generation technology is used. Flash steam technology causes the highest emission because geothermal steam contains a sizable amount of noncondensable gases that escape condensation at the cooling tower into the air. The binary cycle power and heat generation have no emissions because its closed-loop design prevents the geothermal resources from escaping into the air. The emission rate of EGS technology depends on whether it adopts a closed-loop design.

Geothermal plants may change ecosystems and wildlife habitats and reduce species diversity and community composition. Most importantly, the environmental impacts of the large-scale expansion of geothermal plants are still poorly studied and understood on a global scale.

Figure 6.12 shows a sample of the emissions of various geothermal plants and their uses. This indicates that power generation using flash steam has significantly more emissions, CHP and EGS have much fewer emissions, and the HeatORC has virtually no emissions.

5. Well-Drilling Challenges

Rising geothermal power generation and heat use require increased well drilling. According to a study, approximately 1100 geothermal wells were drilled between 2015 and 2020 for power generation, approximately 130 wells will be drilled each year by 2030 for announced geothermal projects, and another 400 wells need to be drilled for unannounced geothermal projects in the same period to meet the countries' renewable targets (Fig. 6.13).

The cost of drilling a geothermal well accounts for approximately 35–60% of the total investment of a geothermal project. This is much more costly than comparable drilling costs for oil extraction because geothermal drilling faces much harsher geological conditions than drilling oil wells. While drilling oil wells typically deals with softer sedimentary rock, geothermal well drilling needs to penetrate hard and abrasive volcanic rocks such as granite or quartzite, which

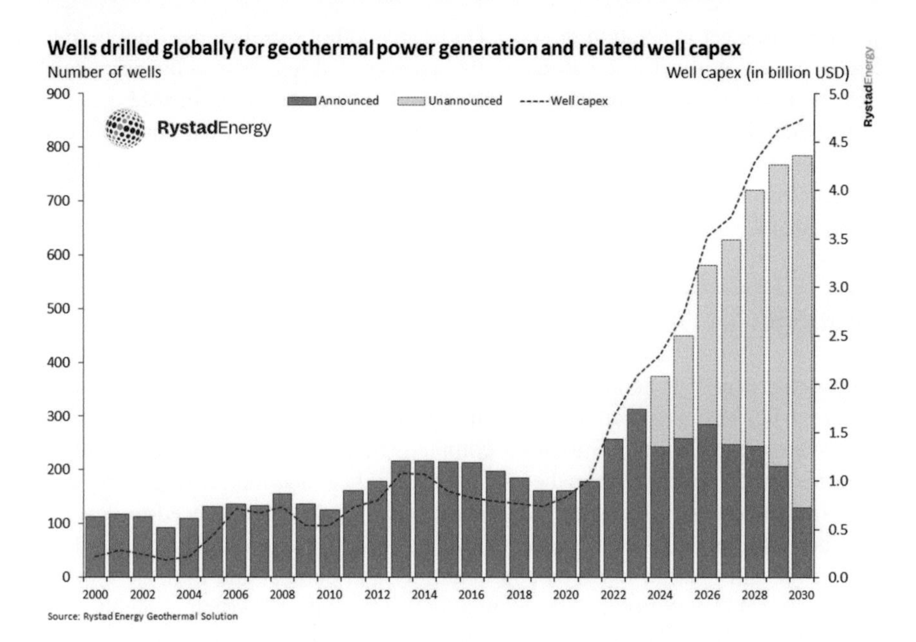

Fig. 6.13 Geothermal well-drilling dynamics (1990–2030). *Image Credit* Rystad Energy, 2023

wear down drill bits much faster. The significantly more frequent replacement of drill bits increases the cost of well drilling and geothermal projects. The costliest challenge of geothermal well drilling is "circulation loss," i.e., the loss of drilling fluid to pores or fractures in the dry rock being drilled. This lost fluid circulation represents approximately 10% of the total costs of a geothermal project and often leads to abandonment of the well during drilling and outputs the geothermal project into economic difficulty.

6. Dry Rock Challenges

The current geothermal industry is dominated by regions with abundant, economically accessible high-temperature hydrothermal resources. Most active geothermal power plants have wells at depths of 2–3 km (~2 miles) and reservoir temperatures of more than 200 °C (392 °F). This high-temperature requirement of geothermal power generation makes the well location factor a bottleneck for the scalability of geothermal power and heat plants. However, only a few countries are endowed with high-temperature reservoirs located at these depths.

Although there are vast amounts of geothermal resources of heat in the hot rock under the surface of the Earth, other geothermal resources of water indispensable for geothermal power generation are often missing. The lack of hot water and steam and reduced amounts of underground water are major challenges for new geothermal power generation projects. Underground hot rock is mostly not permeated by water. Without water stored in the underground reservoir in the permeated rock, there is no natural medium to bring high heat energy to the surface.

Another similar challenge particularly impacts aging geothermal power plants. After years of geothermal power generation, these power plants face diminishing amounts of underground water, which causes reduced well pressure and reduced geothermal steam for power generation.

Solutions

Geothermal energy generation and consumption play a crucial role in the energy transition because geothermal power and heat are reliable and clean renewable energy sources that can be used to complement other intermittent renewables. Maximizing the benefits of geothermal power and heat and minimizing their challenges have become increasingly important because we urgently need to develop innovative technologies to harness renewable energy sources, including geothermal power and heat, to reach carbon neutrality and combat and mitigate climate change. To achieve this objective and to allow large-scale expansion of geothermal energy generation and consumption, every innovative solution is needed to address the challenges to geothermal power and heat generation technologies, including innovative political, economic, and technological solutions.

A. Sustainability of Existing Geothermal Plants

Geothermal power generating projects require land and water. This is especially true for countries with a long history of geothermal power generation, such as the U.S., where a substantial number of dry steam power plants are facing a significant loss of water and loss of related well pressure.

To address the issues of water loss and the related geothermal pressure loss caused by steam evaporation, it is important to reinject steam condensate from the cooling process into the geothermal wells. However, this process cannot prevent a vast amount of steam from escaping into the air. For example, accelerated geothermal power development at The Geysers caused severe steam pressure reduction in the reservoir and power generation by 1998.

To sustain steam production and reservoir pressure, geothermal power plants need a large amount of water to be used for injection. In 1990, Lake County wastewater from the southeastern regional collection system was identified as the preferred source of water for injection. Since 2003, this project has supplied approximately 11 million gallons per day of tertiary treated wastewater to replenish The Geysers geothermal reservoir. Wastewater is treated in three phases—primary (solid removal), secondary (bacterial decomposition), and tertiary (extra filtration) to ensure a high safety level of effluence (water to be released in the natural water system).

Other aging geothermal power plants that experience this sustainability issue can learn from this experience as an effective solution to revitalize their geothermal power and heat generation.

B. Enhanced and Advanced Geothermal Systems

To harness the vast geothermal resources from subsurface dry hot rock that is available anywhere on Earth, it is vital to support and conduct extensive R&D of new geothermal systems such as enhanced or engineered geothermal systems (EGS) or advanced geothermal systems (AGS). These innovative geothermal technologies are aimed at making geothermal more accessible and extractable. Although both EGS and AGS can harness the vast geothermal resources in dry hot rock anywhere on Earth, there are substantial technological differences between them, which make them significantly different in their respective environmental impacts (Fig. 6.14).

	Enhanced (EGS)	**Advanced (AGS)**
Science & Technology	Filling high-pressure water adds water and permeability to the hot rock allowing water to flow through fractured hot rock (artificial geothermal reservoir) and such heated water to turn the turbine/generator to generate power.	Using the working fluid (water or Supercritical CO2[1]) in a coaxial closed U-loop instead of flowing directly into the fractured hot rock allows such a heated working fluid to turn the turbine/generator to generate power.
Open/Closed Loop	Open	Closed
Location Restriction	No	No
Invention	One of the latest types of geothermal power technology, with several R&D and demonstration projects.	One of the latest types of geothermal power & heat generation technology, largely in the R&D and conception stage.
Application	Still at the R&D stage but is expected to get momentum in the future once the high-cost barrier is overcome.	The first commercial Eavor Loop commercial project broke ground in Bavaria, Germany, in October 2022.
Requirement	Need additional R&D in enhancing the dry, hot, and impermeable rock with water and permeability.	A multilateral Eavor-Loop with two laterals, sealed with Rock-Pipe.
Efficiency	It generates power and heat from the dry and impermeable rock without a natural aquifer with an artificial aquifer.	It generates power and heat from the dry and impermeable rock without a natural aquifer with a sealed working-fluid loop.
Sustainability, Environmenta	Less friendly to the environment, with subsurface environmental and seismic impacts because of the need for fracking.	Most friendly to the environment, without the need for fracking and the associated seismic impact.

Fig. 6.14 Comparison of EGS and AGS

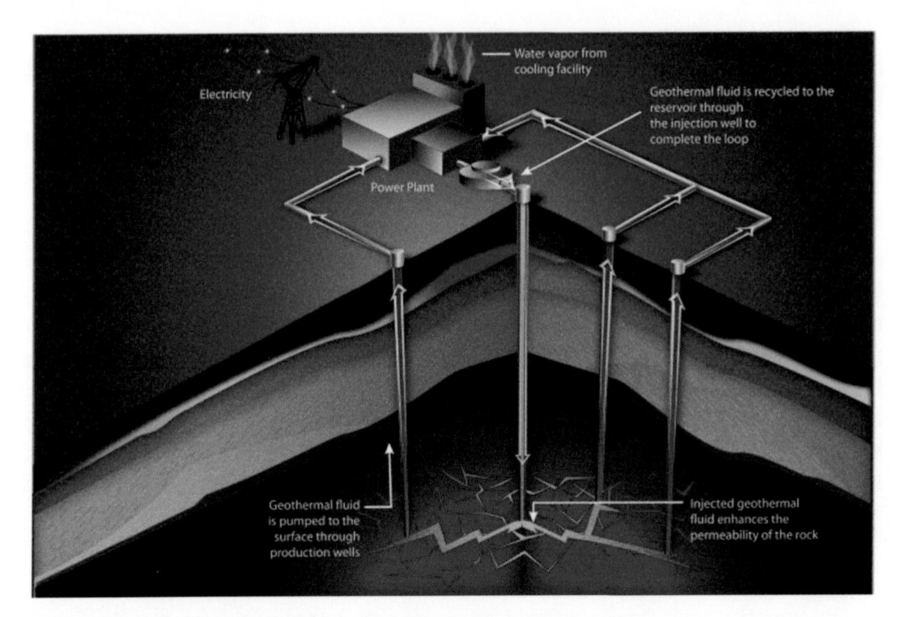

Fig. 6.15 Enhanced geothermal systems. *Image Credit* U.S. Department of Energy

1. Enhanced Geothermal Systems

Enhanced (or Engineered) Geothermal Systems (EGS) are considered an enhanced or engineered solution to the geothermal challenge for most areas on Earth where hot rock exists at an economically accessible depth under the surface but there is no subsurface aquifer or permeability. The EGS is designed to address this location restriction of conventional geothermal power-generating technologies. In addition to drilling a well into the rock, it contains more innovative features, such as creating an artificial subsurface geothermal reservoir and allowing geothermal power generation at locations devoid of natural geothermal reservoirs (Fig. 6.15).

However, EGSs achieve these results by injecting high-pressure water into hot rock, fracturing it, and forcing the water through the fractured hot rock. This process makes EGS an open-loop technology that bears environmental impacts. Although fracturing creates the needed hydrothermal resources through permeable rock for geothermal power and heat generation, injecting a large amount of pressurized water into the Earth's crust to open fissures is an invasive approach that has been found to be positively correlated with increased seismic activity.

Intensified EGS geothermal operations may have the risk of causing serious geological consequences such as seismicity, subsidence, and landslides. Increased deployment of geothermal power and heat generation, especially water reinjection, is associated with increased seismic risks. Studies on the levels of seismic activity at geothermal energy harnessing sites show how the levels vary depending on the activities of the geothermal plants.

Box 6.1 shows the results of a case study on the correlation between intensified geothermal operations and increased seismic activities.

Box 6.1 A case study

A case study was conducted at an active geothermal plant at the southern end of the San Andreas fault line. Its observations showed evidence of such seismic risks associated with active geothermal operations.

Before the geothermal power plant was put into full operation in 2001, the number of seismic activities in this region was extremely low, but since then, geothermal operations and seismicity have both grown. This information shows that geothermal operations and seismic activities are positively correlated.

Certainly, there is uncertainty about the secondary Earthquakes that often occur after an initial Earthquake, which requires more research to determine whether repeated Earthquakes are all associated with human involvement or can some of them be attributed to geothermal energy plants. However, there is a strong correlation between seismic activity and the amount of geothermal activity happening at that time. This correlation is especially obvious between the fluid injection amounts and the net extraction numbers, which indicates that the changes in ground-level fluid volume are causing seismicity.

While most of these Earthquakes have been small seismic activities, there is a real concern about the effect of a larger Earthquake as a result of more aggressive geothermal power and heat generation. There is especially a concern for the San Andreas fault, which has a history of erupting destructive, magnitude eight earthquakes in the surrounding area, which might be massively devastating for the communities all around the area if this correlation between seismic activity and geothermal operations proves true.

2. Advanced Geothermal Systems

Similar to EGS technology, Advanced Geothermal Systems (AGS) technology is specifically engineered to overcome the limitations, particularly those related to geographical constraints, inherent in conventional geothermal power generation technologies such as dry steam, flush steam, and binary geothermal power generation technologies.

However, AGS employs advanced hydrothermal techniques to circumvent the environmental and seismic impacts associated with EGS. AGS does not involve the direct injection of high-pressure water into hot rock formations, nor does it induce

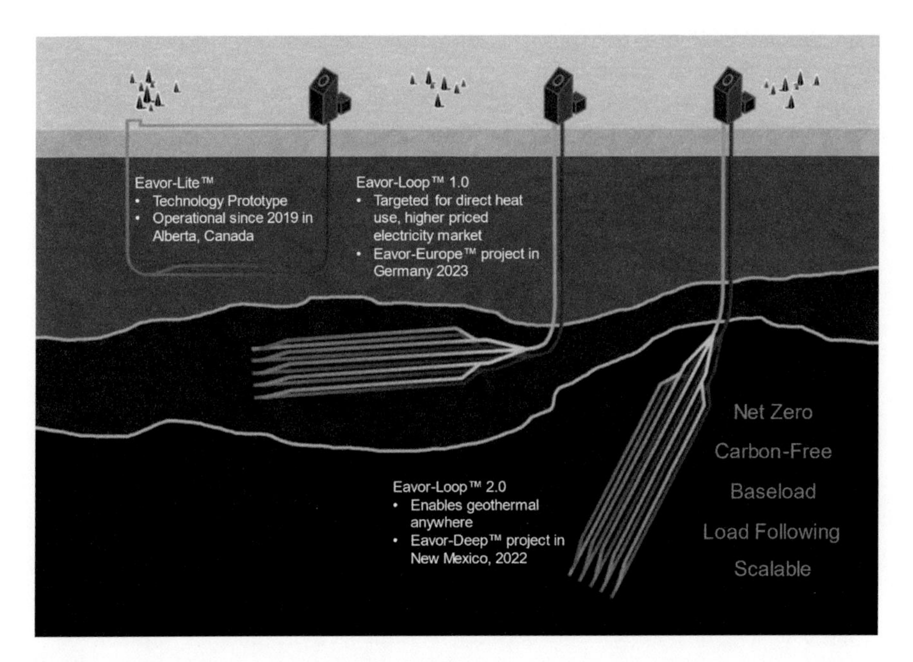

Fig. 6.16 Advanced geothermal systems: eavor loop. *Image Credit* Eavor

rock fracturing or the creation of artificial hydrothermal reservoirs. Additionally, AGS does not rely on the extraction of convective heat from such constructed reservoirs (Fig. 6.16).

Instead, it uses a closed loop, fills it with a working fluid that does not have direct contact with hot rock, and extracts the conductive heat from the working fluid in the closed loop. This closed-loop technology is designed to avoid the seismic and environmental risks of EGSs. As a whole, the no-aquifer, no-fracking and closed-loop features of AGS technology address both the location restriction and high water use issues of traditional geothermal power generation and the environmental and seismic issues of EGSs, such as earthquakes, high water use, produced brine or solids, and aquifer contamination. These features make AGS technology a sustainable geothermal power generation technology.

However, there is a concern regarding the AGS designs' ability to conduct heat at a fast enough rate through the rock. To ensure that the closed-loop heat exchanger optimally harnesses the geothermal heat from hot rock, AGS encompasses advanced features, such as inserting a multilateral closed loop with two laterals and sealing the loop using RockPipe. This design creates a natural upstream geothermal flow.

The insertion of the multilateral closed loop with two laterals can increase the geothermal fluid production by providing additional pathways for the fluid to flow through the rock formations because of enhanced reservoir connectivity and improved wellbore stability. The two laterals can help increase the connectivity between the various parts of the closed-loop geothermal system, allowing for a more efficient and effective transfer of heat from the reservoir to the geothermal fluid. The rock pipe used to seal the loop can help to stabilize the wellbore by reducing the risk of collapse or other issues that could impact the well's performance. AGS systems can have a lower environmental impact than traditional geothermal systems because retaining the working fluid within the system not only limits the amount of drilling required but also prevents it from encroaching on the hot rock, reducing the environmental impact of geothermal power generation. In addition, AGS can also be designed as a power and heat coproduction project, which uses the waste heat of power generation to supply the local area with geothermal heating and can achieve a high efficiency rate of more than 80%.

3. Status of EGS and AGS

Although the commercialization of these innovative geothermal technologies may not be fully viable soon, successful advances in EGS and AGS research, development, and demonstration projects are expected to provide reasonable opportunities for new geothermal technologies to bear fruition in the future and spur a giant surge in geothermal well drilling and geothermal energy expansion in many countries.

In contrast to conventional hydrothermal systems that use open-loop systems to naturally extract convective geothermal energy directly from dry steam or indirectly from flashed hot water to the surface, closed-loop geothermal systems such as binary cycle systems and advanced geothermal systems (AGS) rely on the conductive heat of hot rock to transfer heat to a working fluid circulated within a closed-loop well (binary cycle) or, in more recent designs, through underground loops (AGS).

The AGS underground loop design relies on the thermosiphon effect, where cool water sinks on one side while hot water rises on the other, without the need for a pump. Similar to binary cycle systems, AGS systems have a reduced environmental footprint because no working fluid leaves the sealed closed loop, and the used fluid is directly recycled to the ground. Because AGS systems do not require large volumes of subsurface water or permeability, they have the potential to be built and operate in a greater range of locations.

C. Geothermal Direct Use

Geothermal direct use (GDU) applications, which involve utilizing underground thermal resources directly, i.e., without their conversion to electrical power, can be considered environmentally friendly and sustainable solutions to the challenges of geothermal energy technologies. GDU includes applications such as space heating, cooling, and hot water supply.

There are several advantages to direct geothermal use over geothermal power generation. One of the most significant advantages is its efficiency. Different from geothermal power generation, which involves several energy conversion steps with energy losses, the direct use of geothermal energy for heating applications involves fewer energy conversion steps and therefore has a higher roundtrip efficiency. Moreover, direct geothermal use is often more cost-effective than geothermal power generation because it requires less infrastructure and equipment, making it less expensive to construct and operate. In contrast, geothermal power generation facilities require expensive drilling and piping systems, as well as turbines and generators, which increases their capital costs.

Furthermore, direct geothermal use can provide a consistent and reliable source of heat for applications such as space heating, which is not affected by weather or other environmental factors. In contrast, geothermal power generation can be affected by variations in geothermal resource availability, such as changes in reservoir temperature or water flow rate. Developing geothermal direct use in tandem with geothermal power generation can help to maximize the potential of geothermal resources. While geothermal power generation is crucial for providing renewable power, direct geothermal use can provide low-carbon heating and cooling for buildings, which accounts for a significant portion of global energy consumption.

District heating systems that provide heat to entire neighborhoods or cities are one of the geothermal direct use applications. Reykjavik, Iceland, for example, has one of the world's largest geothermal district heating systems, with approximately 90% of the city's buildings being heated by geothermal energy. Another example is the Oregon Institute of Technology's geothermal heating and cooling system, which uses geothermal water from a nearby well to provide heating and cooling to campus buildings.

Geothermal direct use has several advantages over power generation, including lower capital costs and higher efficiency. For example, a geothermal district heating system can achieve a roundtrip efficiency of up to 92%, while geothermal power generation typically has an efficiency of approximately 10–23%. Additionally, direct use systems require less infrastructure and equipment, making them less expensive to construct and operate. Geothermal direct use has several advantages over power generation and developing it alongside power generation can help to maximize the potential of geothermal resources. Geothermal direct use can provide low-carbon heating and cooling for buildings, and district heating systems can provide heat to entire communities. With continued development and investment, geothermal energy has the potential to become an even more significant contributor to global energy production and decarbonization efforts.

Developing geothermal direct use alongside geothermal power generation can help to unlock the full potential of geothermal resources. For example, in the U.S., the Department of Energy's Geothermal Technologies Office is working to promote both geothermal power generation and direct use. The office has funded projects focused on developing geothermal district heating systems and exploring new direct use applications, such as using geothermal energy for industrial processes.

4. Use of a Heat Pump

A heat pump is a geothermal technology that uses the constant temperature of the Earth to transfer heat from one place to another, either for space heating or cooling. Heat pumps operate on the principle of refrigeration and work by using a refrigerant to transfer heat from a cold environment to a warm environment, and vice versa. In space heating, heat pumps extract heat from the ground or groundwater and use it to warm indoor spaces, while in cooling, they absorb heat from indoor spaces and release it into the ground or groundwater.

One of the main advantages of heat pumps is their energy efficiency. Since they transfer heat rather than generate it, heat pumps can provide up to four times as much heating or cooling energy as the electrical energy they consume. Heat pumps also have low greenhouse gas emissions, making them a sustainable heating and cooling option. Heat pumps are a geothermal technology that utilizes the constant temperature of the Earth to transfer heat for space heating and cooling. The technology is energy-efficient and sustainable and has been widely adopted in various parts of the world, including Sweden, Finland, the U.S., and China. The increasing popularity of heat pumps highlights their importance in reducing greenhouse gas emissions and transitioning to a low-carbon future.

Heat pumps have been used for space heating and cooling in various parts of the world. In Sweden, for example, approximately 80% of new homes are equipped with heat pumps for space heating, and the technology has been widely adopted in the country's district heating systems. In Finland, heat pumps are used for both space heating and cooling, with over 80% of new homes using the technology.

In the U.S., heat pumps have been used for both residential and commercial heating and cooling since the 1940s, with the technology becoming increasingly popular in recent years. According to the Department of Energy, heat pumps accounted for approximately 50% of new heating systems installed in the U.S. in 2020.

Heat pumps have also become a popular heating and cooling technology in China, particularly in the northern part of the country where winters are cold. According to the International Energy Agency, China is the world's largest market for heat pumps, with over 200 million units sold in 2018 alone.

Summary

Geothermal power is a renewable energy generation technology that is used to generate power and heat by mechanically converting hydrothermal energy into electric power. This technology differs from power generation technologies using fossil fuels by using free and clean hydrothermal energy instead of burning fossil fuels to generate power and heat. Therefore, geothermal power is almost as safe and clean as solar power, wind power, and hydropower but much safer and cleaner than power from burning fossil fuels. At the same time, although it is not as clean as solar, wind, and hydropower, it has great potential to offset most

CO_2 emissions from fossil fuel-fired power generation and meet the global Net Zero target. For more than 20 years, government support in many countries has helped geothermal power and heat manufacturers produce more efficient geothermal technologies, such as binary cycle and EGS power plants, through R&D and incentivized geothermal power plant operators to deploy geothermal power technologies. However, geothermal power generation still faces many challenges, especially in EGS expansion, and urgently needs solutions for higher penetration to help meet the global Net Zero goal by 2050.

In the current geothermal power market, large geothermal power plants mainly depend on accessible hydrothermal resources and conventional dry steam and flash steam technologies but are facing challenges such as decreasing hydrothermal resources and increasing environmental challenges such as water shortages.

Geothermal innovations such as the binary cycle, EGS, and AGS provide potential solutions to the challenges of conventional self-flowing steam-based geothermal technologies. However, binary cycle power plants require much more investment and are mostly small geothermal power plants, and EGS and AGS are largely still in the experimental stage. To address the challenges of geothermal power and heat generation, more government support for the R&D of geothermal technological, economic and political innovations is necessary. To reduce the high investment and maintenance cost of geothermal power, more proactive measures need to be taken.

Examples include working out stringent policies on assessing the environmental and ecological impacts of geothermal power generation, educating consumers on price comparison with fossil fuel-fired power in terms of avoided fuel cost and external cost, providing government funding for R&D of closed-loop geothermal technologies such as binary cycle and EGS, continued deployment of geothermal power and heat coproduction plants and digitization and AI technology in smart geothermal well drilling and power generation, expanding geothermal power generation, streamlining the review process to reduce red tape and delays and improve efficiency in permitting and licensing, and providing geothermal power investors with financial incentives or commercial loan programs.

Activities

Further Readings

1. Geothermal Energy Factsheet. https://css.umich.edu/publications/factsheets/energy/geothermal-energy-factsheet
2. Geothermal Energy Challenges: Understanding What Holds Us Back from Tapping into This Abundant Energy Source. https://www.rpsgroup.com/insights/energy/geothermal-energy-challenges-what-holds-us-back-from-tapping-into-this-abundant-energy-source/
3. 2021 U.S. Geothermal Power Production and District Heating Market Report. https://www.nrel.gov/docs/fy21osti/78291.pdf

4. New NREL Report Details Current State and Vast Future Potential of U.S. Geothermal Power and Heat. https://www.nrel.gov/news/press/2021/new-nrel-report-details-current-state-vast-future-potential-us-geothermal-power-heat.html
5. Geothermal energy holds great potential, but technical and regulatory challenges must be overcome. https://www.drillingcontractor.org/geothermal-energy-holds-great-potential-but-technical-regulatory-challenges-must-be-overcome-58226
6. The pros and cons of enhanced geothermal energy systems. https://yaleclimateconnections.org/2020/02/the-pros-and-cons-of-enhanced-geothermal-energy-systems/
7. Large-Scale Enhanced Geothermal System Trial Successfully Completed. https://www.powermag.com/large-scale-enhanced-geothermal-system-trial-successfully-completed/
8. The Rapidly Evolving Enhanced Geothermal Systems Market. https://www.sdcexec.com/sustainability/clean-energy/article/22118938/bis-research-the-rapidly-evolving-enhanced-geothermal-systems-market
9. Supporting Energy Transition by Turning Geothermal Waste Heat into Baseload Energy. https://jpt.spe.org/twa/supporting-energy-transition-by-turning-geothermal-waste-heat-into-baseload-energy
10. An Overview of Geothermal Resources. https://www.thinkgeoenergy.com/geothermal/an-overview-of-geothermal-resources/
11. Closed-Loop Geothermal Technology for a 24/7 Carbon-free and Secure Energy Future. https://www.eavor.com/technology/
12. Geothermal Everywhere: A New Path for American Renewable Energy Leadership. https://innovationfrontier.org/geothermal-everywhere-a-new-path-for-american-renewable-energy-leadership/

Closed Questions

1. Which of the following is geothermal power?
   (a) District heating using geothermal resources
   (b) Using underground hot water to heat buildings, grow plants in greenhouses, and dehydrate onions and garlic
   (c) Using hydrothermal resources to generate electric energy
   (d) Snow melting with ground-coupled heat pumps
   (e) All the above.
2. Which of the following is not currently used for geothermal power generation?
   (a) Flash steam technology
   (b) Dry steam technology
   (c) Binary cycle technology
   (d) Enhanced geothermal systems
   (e) Advanced geothermal systems.

3. Which of the following is the oldest geothermal power generation technology?
   (a) Flash steam technology
   (b) Advanced geothermal systems
   (c) Binary cycle technology
   (d) Enhanced geothermal systems
   (e) Dry steam technology.
4. Which of the following is not a major difference between the binary cycle and the earlier dry and flash steam technologies?
   (a) Binary cycle technology is a more advanced geothermal power technology
   (b) It can be used to generate power anywhere on Earth, whereas earlier technologies do not work without natural hydrothermal resources
   (c) It uses a secondary fluid in the binary cycle to capture heat from geothermal fluids
   (d) With binary cycle technology, hot geothermal fluid is used to heat a secondary fluid with a lower boiling point, such as isobutane or pentane
   (e) The secondary fluid vaporizes and drives a turbine to generate power.
5. What is the common benefit of both EGS and AGS power generation technologies?
   (a) Both can be used to generate power without extensive well drilling
   (b) Both can be used to generate power without the use of substantial amounts of water
   (c) Both can be used to generate power anywhere on Earth
   (d) Both can generate power without risks of triggering earthquakes
   (e) All the above.
6. What is the most significant benefit of AGS technology?
   (a) It is an open-loop power generation system
   (b) It does not require well drilling
   (c) It is the most mature geothermal power generation technology that does not need further innovation
   (d) It is the cleanest geothermal power generation technology that can be used anywhere on Earth
   (e) All the above.
7. Which of the following is the largest hurdle to traditional geothermal power technologies?
   (a) They cause frequent earthquakes
   (b) Their operation requires injecting substantial amounts of water from the beginning
   (c) They are only working at rare locations with ample hydrothermal resources
   (d) They require high fuel costs
   (e) All the above.
8. Which of the following is a challenge for both EGS and AGS?
   (a) Both cause frequent earthquakes
   (b) Both are emerging geothermal power generation technologies that need further innovative R&D

   (c) Both have significant environmental impacts
   (d) Both require ample hydrothermal resources
   (e) All the above.
9. Which of the following technologies is most related to seismic and environmental risks?
   (a) Dry steam technology
   (b) Flash steam technology
   (c) Binary cycle technology
   (d) EGS
   (e) AGS.
10. Which of the following is a shared feature of the binary cycle and AGS technologies?
   (a) Utilizing lower temperature geothermal resources that are not suitable for traditional dry steam or flash steam power plants
   (b) Involving the use of a secondary fluid, which is used to transfer the heat from the geothermal source to the power generation system
   (c) Involving a heat exchanger where the heat from the geothermal source is transferred to the secondary fluid
   (d) Being applicable anywhere around the world
   (e) Being able to generate power in a reliable and sustainable manner.
11. Which of the following is not a solution for the expansion of geothermal energy?
   (a) Developing applications such as space heating, cooling, and hot water supply
   (b) Further upscaling and improving the efficiency of binary cycle geothermal technology
   (c) Providing government financial support for R&D of EGS and AGS
   (d) Upscaling district heating using geothermal resources in residential and business areas
   (e) Ignoring all the above.
12. What is geothermal direct use and how does it differ from geothermal power generation?
   (a) Geothermal direct use involves converting geothermal energy into electricity, while geothermal power generation utilizes heat directly for various applications
   (b) Geothermal direct use and geothermal power generation are the same thing
   (c) Geothermal direct use involves extracting geothermal fluids for industrial processes, while geothermal power generation uses geothermal steam to generate electricity
   (d) Geothermal direct use involves harnessing geothermal energy for space heating and cooling directly, while geothermal power generation produces electricity
   (e) None of the above.

13. What are the advantages of geothermal direct use compared to other forms of heating and cooling?
    (a) Higher costs and lower efficiency compared to fossil fuels
    (b) Limited availability of resources compared to solar energy
    (c) Reliance on fossil fuels compared to wind energy
    (d) Unlimited availability of resources compared to all other energy sources
    (e) Lower costs and higher efficiency compared to traditional heating and cooling methods.
14. What are some notable examples of successful geothermal direct use projects around the world?
    (a) Reykjavik, Iceland‘s geothermal district heating system
    (b) The Geysers geothermal power plant
    (c) Nevada Geothermal Power Plant
    (d) Larderello Geothermal Power Plant
    (e) All the above.
15. How does direct geothermal use contribute to reducing greenhouse gas emissions?
    (a) By increasing reliance on fossil fuels
    (b) By emitting enormous amounts of carbon dioxide
    (c) By reducing the need for combustion-based heating systems
    (d) By causing potential environmental concerns
    (e) None of the above.
16. What is not a viable solution to expand the deployment of geothermal energy technologies?
    (a) By investing in R&D of geothermal resources and innovative geothermal energy technologies
    (b) By providing government support and policy to encourage geothermal energy projects
    (c) By encouraging international collaboration and knowledge exchange
    (d) By increasing reliance on traditional geothermal power technologies despite the lack of abundant underground hydrothermal resources
    (e) By simplifying and streamlining the permitting and regulatory processes for geothermal projects.
17. What are the economic benefits of implementing geothermal direct use systems?
    (a) Higher operating costs
    (b) Increased reliance on imported energy sources
    (c) Job creation and local economic development
    (d) Reduced price stability
    (e) None of the above.
18. What are the environmental considerations and impacts associated with geothermal power generation and direct use applications?
    (a) Air pollution and greenhouse gas emissions

(b) Water usage and potential depletion of geothermal reservoirs
(c) Land disruption and habitat loss
(d) All the above
(e) None of the above.

19. What policies and incentives exist to promote the adoption of geothermal direct use technologies?
    (a) No government support or incentives are available
    (b) Limited funding for research and development
    (c) Government grants, tax incentives, and feed-in tariffs
    (d) Stringent regulations prohibiting geothermal direct use.
    (e) None of the above.
20. What is the difference between Enhanced Geothermal Systems (EGS) and Advanced Geothermal Systems (AGS)?
    (a) EGS focuses on extracting heat from artificial reservoirs in fracked dry rock, while AGS focuses on extracting heat from closed loops in dry rock using advanced drilling techniques
    (b) EGS and AGS are two different terms used to describe the same geothermal technology
    (c) EGS refers to geothermal systems used for electricity generation, while AGS refers to systems used for direct heating and cooling applications
    (d) EGS uses conventional drilling techniques, while AGS involves using geothermal fluids directly from hot springs or geysers
    (e) EGS is designed to extract heat in natural reservoirs, while AGS is designed to extract heat in artificial reservoirs.

Open Questions

1. What is geothermal power? How does it differ from power generation using fossil fuels or other renewable energy sources?
2. What is the binary cycle? What is the main difference between binary cycle geothermal power technology and dry steam and flash steam geothermal power technologies? Why is the binary cycle technology friendlier to the environment than the latter technologies?
3. What are the requirements for various types of geothermal power? What challenges do these requirements pose?
4. Why does geothermal power installation have high upfront costs? What can be done to help reduce geothermal power costs?
5. What potential challenges do large utility-scale geothermal power plants pose to other grid-connected renewable energy generation? What potential solutions are there to address these challenges?
6. Is there a high potential for geothermal power to be curtailed? Explain why you think so.
7. What role does geothermal power play in the power grid?

8. What can be done to revitalize aging geothermal wells for power generation?
9. What is the importance of water for geothermal power and steam generation?
10. What are the crucial factors of repurposing old dams for geothermal power generation?
11. What type of geothermal power technology is most friendly to the environment?
12. What type of geothermal power technology has the most potential for most countries that do not have accessible hydrothermal resources?
13. What are the differences between open-loop and closed-loop geothermal systems? Which one is more eco-friendly? Why?
14. What are enhanced geothermal systems (EGS)? What is the main difference between EGS and conventional geothermal power technologies?
15. What are the benefits and challenges of EGS? Why is the R&D of EGS important for future geothermal power deployment in the action against climate change and global warming?
16. What are advanced geothermal systems (AGS)? What is the main difference between AGS and EGS?
17. What are the benefits and challenges of EGS? Why is the R&D of AGS important for future geothermal power deployment in the action against climate change and global warming?
18. What are the viable solutions to these challenges?
19. Do both geothermal power and geothermal direct use rely heavily on extensive drilling? Explain your answer with the fundamental differences between them.
20. Of EGS and AGS, which has more in common with the binary cycle geothermal? Explain the similarities and differences between similar technologies.

7 Energy Storage

Introduction

Science and Technology

Energy storage refers to the processes, technologies, or equipment with which energy in a particular form is stored for later use. Energy storage also refers to the processes, technologies, equipment, or devices for converting a form of energy (such as power) that is difficult for economic storage into a different form of energy (such as mechanical energy) at a lower cost for later use.

Energy storage, especially grid-connected energy storage, plays a vital role in renewable energy transformation because robust power generation from intermittent variable renewable energy sources needs robust and powerful energy storage to maintain balance and stability for power generation, transmission, and distribution on the grid. The need for energy storage and flexibility is growing with increasing shares of variable renewable energy (VRE) and phasing out of fossil power plants. Grid stability, grid resilience, and sufficient flexibility options for load-generation balancing will be central to planning and operating renewable energy-dominated grids in the future.

1. Energy Storage Classification

There are several ways energy storage is classified. The following is a list of the main classifications.

(a) *Duration-based classification* distinguishes between energy storage systems that are designed for short-duration applications (such as capacitors or some battery chemistries) and those that can store and discharge energy over a period of hours, days, or months (such as some flow batteries or thermal energy storage systems). Different energy storage technologies ensure renewable energy

P. Yang, *Renewable Energy*, https://doi.org/10.1007/978-3-031-49125-2_7

Flexibility Type	Duration	Issue	Importance for grid operation a. planning
Short term	Sub-second to seconds	System stability	Dynamic stability: Inertia response, voltage a. frequency
	Seconds to minutes	Short-term frequency	Primary a. secondary frequency response
	Minutes to hours	More fluctuations in the supply/demand balance	Real time power market balance
Medium term	Hours to days	Hourly a. daily operation scheduling	Energy supply a. demand balance
Long term	Days to months	Longer-duration VRE surplus or deficit	Energy adequacy scheduling over longer period
	Months to years	Seasonal a. inter-annual VRE availability	Hydro-thermal coordination, adequacy, power system planning (energy over very long durations)

Fig. 7.1 Functionality of energy storage for renewable energy

stability and sustainability in the short, medium, and long terms. Figure 7.1 shows the key role that grid-connected energy storage technologies designed for different durations play in grid operation and planning.

(b) *Scale-based classification* distinguishes between large energy storage systems that serve a grid- or utility-scale system (such as pumped hydro storage) and those that are designed for smaller-scale distributed energy applications (such as residential solar PV + storage systems or residential solar heat storage systems).

(c) *Technology-based classification* is the most common way of presenting energy storage types that distinguishes energy storage systems based on the technologies of energy storage. There are four major technological types of energy storage: mechanical energy storage, electrical energy storage, electrochemical energy storage, and thermal energy storage (Fig. 7.2).

2. Energy Storage Technologies

(a) Mechanical Energy Storage

Mechanical energy storage, also known as kinetic or potential energy storage, currently holds the largest share of global energy storage capacity. It primarily relies

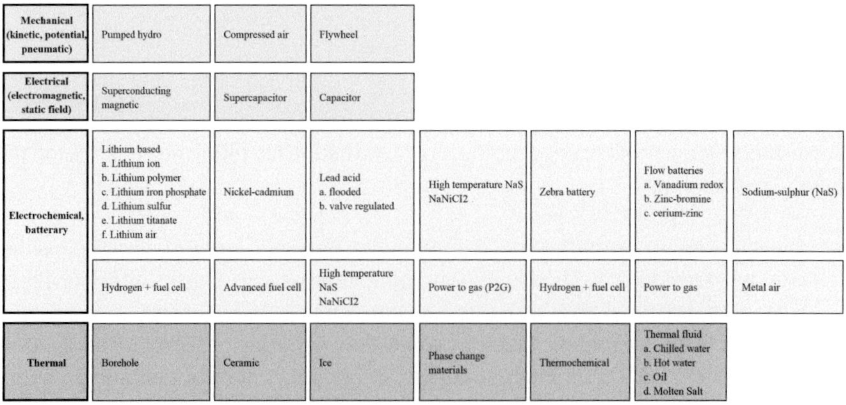

Mechanical (kinetic, potential, pneumatic)	Pumped hydro	Compressed air	Flywheel				
Electrical (electromagnetic, static field)	Superconducting magnetic	Supercapacitor	Capacitor				
Electrochemical, batterary	Lithium based a. Lithium ion b. Lithium polymer c. Lithium iron phosphate d. Lithium sulfur e. Lithium titanate f. Lithium air	Nickel-cadmium	Lead acid a. flooded b. valve regulated	High temperature NaS NaNiCl2	Zebra battery	Flow batteries a. Vanadium redox b. Zinc-bromine c. cerium-zinc	Sodium-sulphur (NaS)
	Hydrogen + fuel cell	Advanced fuel cell	High temperature NaS NaNiCl2	Power to gas (P2G)	Hydrogen + fuel cell	Power to gas	Metal air
Thermal	Borehole	Ceramic	Ice	Phase change materials	Thermochemical	Thermal fluid a. Chilled water b. Hot water c. Oil d. Molten Salt	

Fig. 7.2 Major energy storage technologies

on the conversion of kinetic energy, which can take many forms, such as gravitational, rotational, translational, vibrational, and thermal energy, into other forms of energy for storage and the conversion of stored kinetic energy back into other forms of energy when needed.

These technologies work through complex systems that involve turbines, compressors, and other machinery, utilizing water, air, or heat as the medium for energy transfer and storage. Mechanical energy storage is commonly used in a variety of applications, including power grid stabilization, renewable energy integration, backup power supply, transportation, and industrial and commercial applications. It encompasses several technologies, including pumped hydropower storage, compressed air energy storage, and flywheel energy storage.

Pumped hydropower storage (PHS) is a mechanical energy storage technology that plays a vital role in storing grid power for balancing loads in power systems. It uses surplus renewable energy such as solar PV or wind power that cannot be used during low-demand periods to pump water to a higher-elevation reservoir. The pumped hydro stores the energy in the form of the gravitational potential energy of water. When the power demand is high, the stored water is released from the higher reservoir, flowing back down through hydro turbines to generate power. Important applications of PHS include a power storage medium for various renewable energy storage, ancillary grid services, and storing power for other purposes.

Technically, PHS is similar to hydropower. However, different from hydropower, the flow for PSH units is bidirectional. Although not always, the same equipment is usually used for both generation and pumping. Therefore, the synchronous generator also operates as a motor, and the hydro turbine also operates as a pump. Both components are reversible in their functionality. Some PHS plants, particularly those where the head (i.e., the effective height between the water source and the turbine) is extremely high, may require separate turbines and pumps.

There are two types of PHSs, i.e., open-loop PHS and closed-loop PHS. While open-loop PHS is hydrologically connected to a natural body of water, either a river, lake, or ocean, the water in a closed-loop PSH only moves between the upper and lower reservoirs and does not move to an outside body of water (Fig. 7.3). Because closed-loop PHS does not interact with a natural body of water, it has a less negative environmental impact than open-loop PHS.

Compressed-air energy storage (CAES) systems typically store energy in compressed-air form. This technology uses surplus renewable energy to compress air and stores the compressed air in underground caverns or other large, sealed containers, which can be released later to generate power when needed. CAES systems use three different thermal storage processes: diabatic, adiabatic, or isothermal storage technologies. Each of them has effective applications (Fig. 7.4).

Diabatic or uninsulated CAES systems store compressed air without any significant insulation to prevent energy loss. This is because the primary purpose of these systems is to store energy in the form of compressed air, rather than heat or cold as in thermal energy storage systems. Some diabatic CAES systems use insignificant amounts of insulation or other methods to minimize energy losses during storage

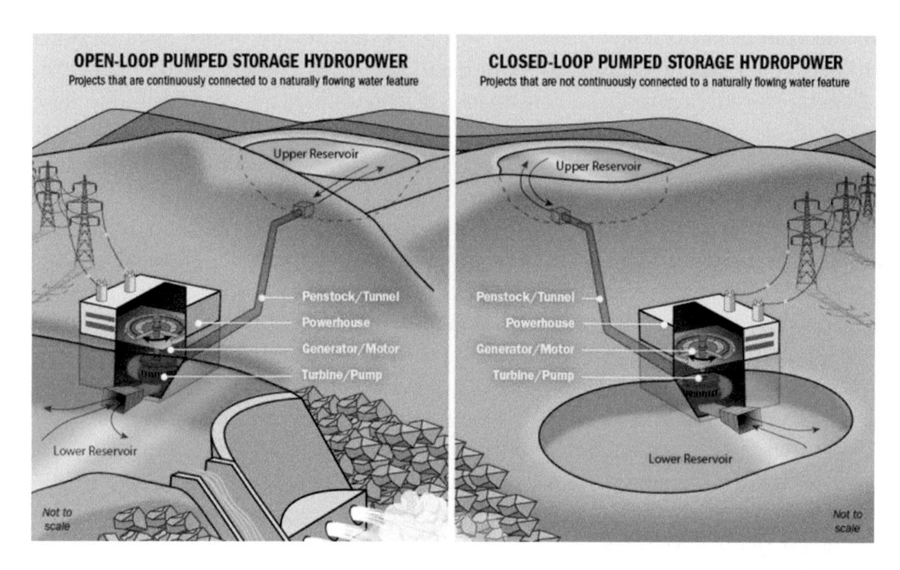

Fig. 7.3 Open-loop PHS versus closed-loop PHS. *Image Credit* U.S. Department of Energy

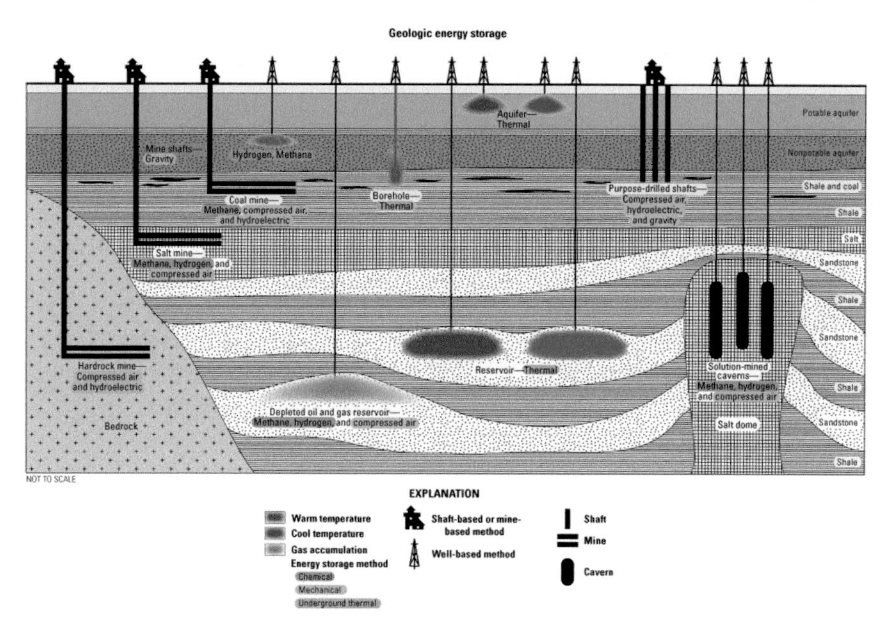

Fig. 7.4 CAES in various underground settings. *Image Credit* U.S. Geological Survey

and discharge. Diabatic CAES is particularly useful for utility-scale energy storage applications. Diabatic CAES systems are usually less expensive than the other two adiabatic or isothermal types, but they are also less efficient.

Adiabatic or insulated CAES systems store compressed air with significant insulation to prevent energy loss. This feature provides higher efficiency because all energy stored in compressed air is fully contained. Adiabatic CAES is particularly useful for generating power quickly, such as in grid stabilization or backup power applications. Because of insulation and no need for additional heating or cooling energy input during the compression and expansion processes, adiabatic CAES systems typically have higher round-trip efficiency than diabatic or isothermal systems. However, they have higher capital costs due to the need for additional components such as thermal management systems, heat exchangers, and thermal storage systems.

Isothermal (meaning constant temperature) CAES systems are designed to maintain a constant temperature during compression and expansion, which requires a significant amount of thermal management through heating and cooling. This design increases the efficiency and response times of CAES systems, making them well suited for applications that require quick and reliable energy delivery, such as microgrids or remote power applications. Isothermal CAES systems are typically more expensive and complex than adiabatic or diabatic CAES systems due to the need for advanced thermal management and the use of specialized equipment. However, they offer higher efficiency and greater flexibility in terms of operating conditions, which allows them to provide a promising option for large-scale energy storage applications.

Flywheel technology uses electric motors to drive a rotating mechanical device called a "flywheel" to spin at a high speed. This technology transforms and stores electric power into kinetic (mechanical) energy in the form of a rotating wheel. To prevent energy loss, a magnetic field is created to maintain the wheel in a frictionless high vacuum environment. With magnetic bearings and high vacuum, flywheels can maintain 97% mechanical efficiency and 85% round trip efficiency (Fig. 7.5).

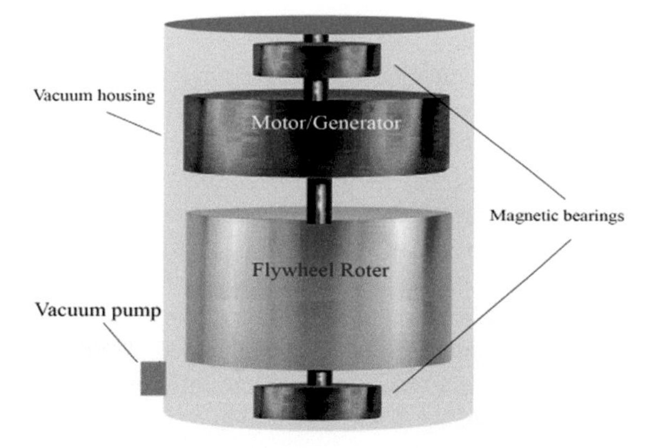

Fig. 7.5 Flywheel

When power is needed, the stored kinetic energy is used to operate the motorized generator to generate power, prevent obstructions in major power systems, and help in the maintenance of mechanical system adjustments. Flywheel technology is primarily used for short-term energy storage and for providing frequency regulation services to the power grid. It is also used in uninterruptible power supply (UPS) systems to provide backup power during power outages. Moreover, flywheels can be used in hybrid systems with other storage technologies, such as batteries, to provide longer-duration energy storage solutions.

(b) Electrochemical Energy Storage

Batteries store electric power electrochemically, which can be discharged for later use. Batteries store power by means of a chemical reaction between the positive and negative electrode materials and the electrolyte. When the battery is being charged, a current is applied to it, which causes a chemical reaction that stores energy in the form of chemical potential. When the battery is discharged, the opposite chemical reaction occurs, which releases the stored energy in the form of electrical power.

There are diverse types of batteries, and these types are named after the chemistries they use, for example, lithium-ion (Li-ion), nickel metal hydride (NiMH), and lead acid. Rechargeable batteries have a chemical composition and one or more electrochemical cells that can be recharged and discharged repeatedly before their performance begins to deteriorate.

The performance and efficiency of Li-ion batteries are much higher than those of other types of batteries because they have a higher energy density per unit weight, which allows them to store more energy in a smaller package. They also have a high charge and discharge cycle life, which allows them to be discharged and recharged for many more cycles than other types of batteries. In addition, Li-ion batteries use fewer toxic materials than other types of batteries, which makes them more environmentally friendly. In contrast, nickel-cadmium and lead acid have lower levels of efficiency due to lower energy density, shorter cycle life, slower charging and discharging speeds, and more environmental and health impacts.

Hydrogen and fuel cells combined are considered a type of electrochemical energy storage in addition to diverse types of batteries. This classification is based on the electrochemical conversion of electric energy into hydrogen through the electrolysis of water, the use of the generated hydrogen as an energy carrier or energy storage, and the electrochemical conversion of the stored hydrogen with oxygen into electricity and water using a fuel cell.

Different from batteries, hydrogen fuel cells do not store hydrogen but need a continued supply of hydrogen from the hydrogen tank or onsite electrolysis. However, hydrogen fuel cells are still considered a type of electrochemical energy storage technology because they are used to convert hydrogen or other fuels into power through a chemical reaction between hydrogen and oxygen. They can be used for stationary power generation or power supply in vehicles.

(c) Electrical Energy Storage

The electrical (also referred to as electromagnetic, electrostatic) type of energy storage includes superconducting magnetic energy storage, supercapacitors, and capacitors.

Superconducting magnetic energy storage (SMES) is an energy storage technology that stores power in the form of a magnetic field created by superconducting coils, which are made of a material that can conduct electricity with zero resistance at extremely low temperatures (typically below 10 K (approximately equal to – 263.15 °C or – 441.67 °F). When a current is passed through the coil, it generates a magnetic field, which is stored in the coil until it is needed. When stored energy is needed, the magnetic field is released, and the energy is converted back to electrical energy. The main advantages of SMES are its high efficiency (95%) or low energy loss during the charging and discharging processes, high power density, and fast response times, which make this energy storage particularly useful for applications of short-term power backup, high power output, and grid voltage stabilization.

A capacitor is an energy storage technology that stores electric power electrostatically in an electric field between two charged plates. It consists of one or more pairs of two conductive plates (called conductors) separated by an insulating or dielectric material (called an insulator). When a voltage (i.e., difference in electric potential measured in volts) is applied across the plates, electric charge builds up on the plates, and an electric field is created between them. The amount of charge that a capacitor can store is determined by its capacitance (i.e., ability to store electric charge), which is measured in farads (F). Capacitors generally have lower density and limited energy storage capacity compared to other energy storage technologies, such as batteries or supercapacitors. However, they can provide high power output and fast charging/discharging rates, which makes them useful for applications that require rapid bursts of energy, such as power quality improvement, energy storage for pulsed power applications, and regenerative braking systems [that recover energy loss when brakes are applied] in electric vehicles. Capacitors are typically used in electronic circuits for functions such as filtering, decoupling, and timing.

Supercapacitors (also known as electrochemical capacitors, ultracapacitors or electric double-layer capacitors) are a type of energy storage technology that can store much more energy than traditional capacitors by using both electrostatic and electrochemical principles. On the one hand, unlike traditional capacitors, which use an insulating material (called a solid dielectric or insulator) to store electric charge, supercapacitors use a double layer of ions at the interface between the electrode and the electrolyte as the source of the electric charge. On the other hand, supercapacitors use a high-surface-area electrode material, such as activated carbon, metal oxides or conductive polymers, to allow many ions to be stored in the double layer, resulting in high capacitance.

Box 7.2 Debate on Classification

It is noteworthy that there is a debate among the scientists on how to classify supercapacitors. The focus of the debate is whether supercapacitors belong to electromagnetic energy storage or electrochemical energy storage.

Indeed, the way supercapacitors store energy through the separation of charges at the electrode–electrolyte interface is based on the principles of electrochemistry. At the same time, the way the energy stored in a supercapacitor is related to the electric field generated by the separated charges is based on the principles of electromagnetism.

The way the high capacitance of supercapacitors is achieved by using high-surface-area electrode materials to increase the number of ions that can be stored in the electric double layer is an electrochemical process. However, the way the energy stored in a supercapacitor is related to the electric field between the separated charges is an electromagnetic or electrostatic phenomenon.

Supercapacitors typically have capacitance values in the range of hundreds or thousands of farads compared to the microfarad range for traditional capacitors. In terms of energy storage, supercapacitors can store much more energy per unit volume or weight than traditional capacitors. They also have a much higher power density and can discharge energy quickly. Supercapacitors fill the void between the regular capacitor and the rechargeable battery. They have a high energy density of all capacitors. Its charge or discharge cycle is shorter than that of other capacitors.

However, supercapacitors have a lower energy density than batteries and are not capable of storing as much energy per unit volume or weight as batteries. Therefore, they are better suited for applications where high-power output is needed for short periods of time, such as electric vehicles, regenerative braking systems, some renewable energy systems, elevators, and cranes, as well as static random-access memory backup (SRAM) in critical control systems, aerospace applications, and medical devices.

(d) Thermal Energy Storage

Thermal energy storage (TES) is a technology or process of storing thermal energy (either heat or cold) in a thermal container or material for later use. TES systems typically include storage tanks using molten salt, oil, water, and phase change materials as storage media that can absorb and release thermal energy.

The process of storing thermal energy involves adding thermal energy to the storage system, typically during periods of low demand or excess supply, and discharging the system by withdrawing thermal energy from it, typically during periods of high demand or supply shortage. It is a key strategy for managing and optimizing the use of intermittent and variable energy from renewable sources such as solar PV, solar heat, and wind power.

TES can be used in a variety of applications, including balancing the intermittent output of renewable energy sources in power generation, space heating and cooling in buildings, and metal casting or cement production in industrial processes. By storing thermal energy during periods of low demand or excess supply and releasing it during periods of high demand or supply shortage, TES can help meet peak demand, improve energy efficiency, and reduce greenhouse gas emissions.

While water has been and remains the most common and widely used medium of thermal energy storage (in the forms of hot water, chilled water, or ice) due to its availability, low cost, and thermal properties, molten salt has become the dominant thermal medium for the latest large-scale, high-temperature (up to 600 °C/ 1112 °F or higher) heat transfer and storage in applications such as concentrated solar power plants.

In addition, oil is also commonly used as a medium of heat transfer and energy storage in industrial processes in boilers, heat exchangers, and several types of machinery. Both molten salt and thermal oil can be used in solar thermal power plants to store heat energy during the day and release it at night to generate electricity. Similarly, oil-filled heaters and radiators can store thermal energy and release it gradually to heat a room.

Phase change materials refer to substances that can change their physical state (known as phase) from either solid to liquid (melting) or vice versa (freezing), vaporization (steaming), or condensation (cooling) when they absorb or discharge energy in terms of heat (known as latent heat) at certain temperatures. Water is the most common example of a phase change material that can store and release thermal energy during the process of heating and cooling as well as melting and freezing. in the forms of ice (below 0 °C or 32 °F), liquid (water between 0 and 100 °C or 32–212 °F), and steam (100 °C/212 °F).

TES systems can be classified based on storage materials, operation temperatures, applications, technologies, and durations.

Based on *storage materials*, three main types of TES systems can be found: sensible heat storage, latent heat storage, and thermochemical storage. Sensible heat storage uses materials such as water, oil, rocks, sand, or molten salt that can store heat in them. Latent heat storage uses phase change materials such as water/ice[1] or paraffin[2] that can absorb and release substantial amounts of

---

[1] Water/ice is a very useful phase change material because ice requires much heat (333.55 J/gram) to melt to water, but water only needs an additional 4.18 J/g to further rise 1 °C. It has been used to store winter cold to cool buildings in summer since at least the time of the Achaemenid Empire.

[2] Paraffin, also known as paraffin wax or petroleum wax, is a flammable, waxy solid distilled from petroleum or shale. It is the most commonly used phase-change material for electronics thermal management due to its wide range of melting temperatures, dependable cycling, and noncorrosive, chemically inert properties. However, its low thermal conductivity coefficient (~0.2 W/m °C) can limit its use in applications that require rapid and efficient energy storage and release.

heat energy during melting and freezing. Thermochemical storage uses materials such as metal hydrides or sorbents that can store heat energy through reversible chemical reactions.

Based on *operating temperatures*, three main types of TES are recognized: low-temperature TES systems, medium-temperature TES systems, and high-temperature TES systems. Low-temperature TES systems (below 100 °C/212 °F) are typically used for space heating and cooling. Medium-temperature TES systems (between 100 and 400 °C or 212 and 752 °F) are used for industrial processes and power generation. High-temperature TES systems (above 400 °C/752 °F) are used for CSP and other high-temperature applications.

From the perspective of *applications*, three main TES types can be identified: TES for residential and commercial buildings, TES for industrial processes, and TES for power generation. While TES systems used for space heating and cooling in residential and commercial buildings are low-temperature systems, TES systems for process heating and cooling in industrial processes are medium-temperature TES systems, and TES systems for grid-scale energy storage in power ice, paraffin or salt hydrates. Thermochemical TES uses chemical reactions to store and release generation are high-temperature systems.

There are five types of TES technologies: water- or oil-based TES, phase change material TES, thermochemical TES, molten salt TES, and underground TES. Water-based TES uses large tanks of hot or chilled water to store and release heat (sensible energy) or cold (latent energy) for later use. Oil-based TES uses large tanks of hot oil to store and release heat for later use. Phase-change material TES uses materials that change phase during charging and discharging, such as heat, such as using metal hydrides. Molten salt TES uses a mixture of salts as a heat transfer and storage medium. Underground TES uses the ground as a storage medium.

Duration-based TES systems include two main categories: diurnal TES and seasonal TES. Diurnal TES is used for daily heat storage, such as for shifting thermal demand on space-heating or cooling or power generation within a day. Seasonal TES is mainly used for seasonal storage, such as storing surplus solar heat during the summer for use during the winter.

Benefits

Energy storage technologies play an increasingly vital role in the energy transition and even more substantial role in the 100% renewable energy future because energy storage addresses the intermittent nature of renewable energy sources, stabilizes power grids, and provides backup power. Each energy storage technology has its own set of benefits and limitations, and the choice of technology depends on factors such as the scale of storage needed, location, cost considerations, and specific use cases. A combination of these technologies can create a diverse and flexible energy storage infrastructure that supports the transition to a more sustainable and reliable renewable energy system.

1. **Pumped Hydropower Storage (PHS)**

Using the mature technologies of hydropower and reservoirs, pumped hydropower storage (PHS) addresses the needs of storing massive excess renewable energy during off-peak hours to so that it is not curtailed or wasted and providing this stored energy during hours of peak demand.

PHS can store enormous amounts (GW) of energy for extended periods, making it suitable for grid-scale applications.

PHS systems have high round-trip efficiency and low energy loss, meaning these energy storage systems can efficiently store and retrieve energy.

PHS systems also have grid-scale flexibility because they can respond quickly to changes in demand and supply, aiding grid stability.

2. **Compressed Air Energy Storage (CAES)**

Using the technology of compressing air and storing it in underground caverns, CAES systems meet the same energy storage needs as those the PHS systems meet. These systems have similar benefits to those of PHS systems. CAES can be scaled up to provide substantial energy storage capacity, making it suitable for grid-level applications. These energy storage systems can achieve relatively high energy efficiency when they are used in combination with advanced adiabatic or isothermal processes.

3. **Flywheels**

Using the technology of storing energy in the rotational motion of a spinning flywheel, these energy storage systems address the rapid response needs of grid regulation and stabilization. Flywheels can respond rapidly to changes in power demand or supply, making them suitable for frequency regulation and grid stabilization. Flywheels have a longer lifespan than some other energy storage technologies because of their mechanical nature.

4. **Thermal Energy Storage (TES)**

TES systems meet the needs of storing heat or cold for later use. This technology can be used in conjunction with solar thermal power plants, HVAC systems, and industrial processes. Storing heat or ice when they are abundant and excessive for later use when heat or ice are needed allows for loss, demand, and cost management of thermal energy. In some applications, the optimization of energy usage allows TES systems to enhance their overall efficiency.

5. **Batteries**

As one of the most popular energy storage technologies that are widely used in portable electronics, electric vehicles (EVs), and grid applications, batteries,

Lithium-ion (Li-ion) batteries in particular, have gained prominence due to their high energy density, efficiency, and decreasing costs. Along with the renewable energy penetration in grid, grid-connected Li-ion batteries can quickly respond to fluctuations in energy demand and supply and help stabilize power grids by providing power during peak demand or absorbing excess energy during low demand. Batteries can store excess energy generated by renewable sources (such as solar and wind) and release it when needed, enabling a smoother integration of renewables into the grid. Batteries can also be used provide backup power during grid outages, ensuring uninterrupted power supply to critical infrastructure, businesses, and homes.

Deployment

A. Energy Storage Technology Deployment

The storage of water in reservoirs is not new. Humans have built sophisticated reservoirs to store and distribute water for irrigation, flood control, and other purposes, especially during droughts, since ancient times. One of the most famous reservoirs, the Dholavira Reservoir, dates to 2600 BCE. However, using reservoirs to store electric power took place much more recently. The first PHS facility was built in the U.S. in 1929 when the Hinds Energy Conversion System was built as a 1.75 MW facility used to store surplus power from a nearby coal-fired power plant by pumping water from a lower reservoir to an upper reservoir.

With the rise of variable renewable energy, the demand for reliable, flexible, and low-carbon energy storage systems has become an increasingly important driver for energy storage deployment. The installed energy storage power capacity grew from 105 GW in 2002 to 174 GW in 2021, by 66% in the last 20 years or on average 3.3% every year. However, the installed storage capacity grew at a significantly slower pace, from 1546 GWh in 2002 to 1656 GWh in 2021 (Fig. 7.6).

Overall, the growth of energy storage technologies in the last two centuries was driven by declining costs, improvements in performance and efficiency, and policy support for energy storage deployment. While PHS remains the dominant technology in large-scale energy storage projects, other technologies, such as Li-ion batteries and flow batteries, are also experiencing rapid development as the global energy storage market continues to grow.

By country, the U.S. leads the world in both the number of energy storage projects and the energy storage capacity. Among the top 10 countries with installed energy storage capacity, the U.S. has more energy storage projects than all other nine countries combined, i.e., 874 operational energy storage projects with 272 GWh of energy storage capacity versus 820 projects and approximately 272 GWh of energy storage capacity, i.e., 40% of the global energy storage of

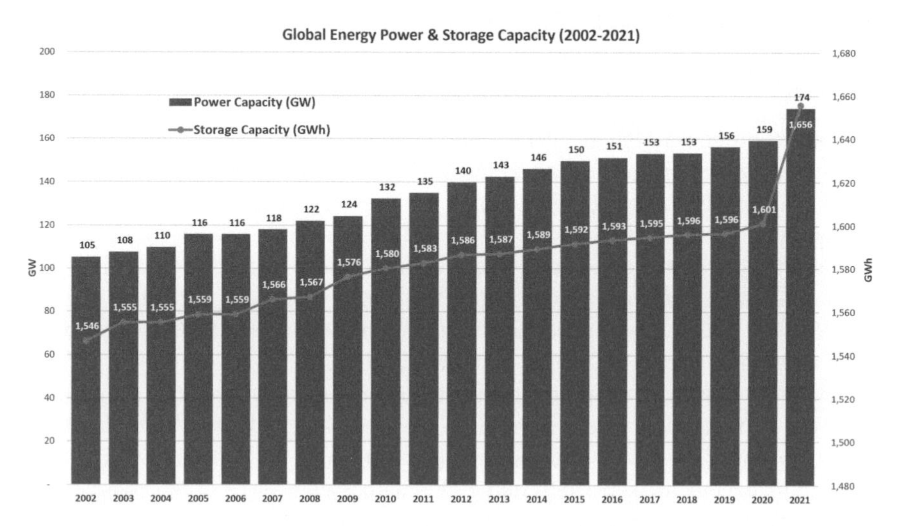

Fig. 7.6 Global energy storage power and storage capacity. *Data Source* DOE Global Energy Storage Database, 2023

683 GWh, although its energy storage rated power (i.e., the installed power generation capacity) of 24 GW is less than that of China (31 GW). It has the world's second largest PHS system—the Bath County Pumped Storage Station, located in Virginia, U.S., which has an installed rated power of 3003 MW and a storage capacity of 31 GWh (Fig. 7.7).

	Number of Projects	Power Capacity (GW)	Storage Capacity (GWh)	Storage Duration (h)
Total	1,694	174	683	3.9
China	90 (5%)	31 (18%)	12 (2%)	0.4
U.S.	874 (52%)	24 (14%)	272 (40%)	11.1
Spain	68 (4%)	9 (5%)	8 (1%)	0.9
Germany	79 (5%)	8 (4%)	46 (7%)	5.9
Italy	54 (3%)	7 (4%)	0.4 (0%)	0.1
India	15 (1%)	7 (4%)	11 (2%)	1.6
Switzerland	24 (1%)	6 (4%)	0.04 (0%)	0
France	27 (2%)	6 (3%)	0.03 (0%)	0
South Korea	58 (3%)	5 (3%)	21 (3%)	4.2
Austria	19 (1%)	5 (3%)	4 (1%)	0.8
Rest of World	386 (23%)	66 (38%)	309 (45%)	4.7

Fig. 7.7 Energy storage by country: top 10 and the rest. *Data Source* DOE Global Energy Storage Database, 2023

B. Energy Storage Technologies

1. Pumped Hydropower Storage

Pumped hydropower storage (PHS, also called "pumped storage hydropower" or "pumped storage"), is the leading member of the mechanical energy storage group, has a long history of dominating energy storage technologies. The first pumped hydropower storage facilities were developed in Switzerland and Italy. They used dual reservoirs with a connecting water conduit and a hydropower turbine generator. Although PHS has a rich history that spans more than a century and has played a key role in power generation, demand management, and grid stability, its rapid expansion started in the mid-twentieth century.

The U.S., Europe, and Japan led the deployment of these projects. Large-scale deployment of pumped storage technology began in the 1960s and 1970s, driven by the expansion of the power grid and the increasing need for grid stability and peak demand management. The 1980s and 1990s witnessed a surge in the construction of pumped storage projects, but this development slowed down in the early 2000s, which was attributed to the liberalization of power markets and the associated difficulty in financing long-term, capital-intensive projects such as pumped storage.

The rise of variable renewable energy sources such as wind and solar power led to a renewed interest in pumped storage in the late 2000s and 2010s. These intermittent renewable energy sources require a balancing function that pumped storage can efficiently perform. Significant technological advancements were made in this period, including adjustable speed technology and ternary pumped storage units (Fig. 7.8).

Due to its technological maturity and long life cycle of 80 or more years, PHS has been the cheapest and most dominant large-capacity grid-connected energy storage technology among all types of energy storage. Its global installed power capacity was approximately 164 GW, and the global storage capacity was approximately 1594 GWh in 2021, which indicates that the installed PHS power capacity accounted for 94% of all installed energy storage power capacity and that its storage capacity accounted for 96% of all installed energy storage capacity.

In recent years, there has been a surge in the development of PHS in China, which accounts for a considerable proportion of the global new pumped hydro projects. As of 2023, the world's largest PHS is the Fengning Pump Storage Power Station in China, with a total installed capacity of 3.6 GW consisting of twelve 300 MW reversible pump-turbine generators and a total energy storage capacity of 40 GWh.

Other countries, such as the U.S., Australia, and Europe, are also investing in new PHS projects. For example, the U.S. has several PHS projects under development, including the proposed Eagle Mountain PHS project in California, which would be one of the largest such projects in the world.

In Australia, the Snowy 2.0 project, which involves the construction of a new PHS facility, is currently under construction. This project is expected to significantly increase Australia's energy storage capacity and help support the integration

Technology	Power Capacity (MW) / Share (%)	Storage Capacity (MWh) / Share (%)	Storage Duration (h)	Projects Number & Share
Pumped hydro storage	164,431 (94.3%)	1,594,397 (96.3%)	9.7	364 (22%)
Compressed air energy storage	722 (0.41%)	33,301 (2%)	46	11 (0.7%)
Flywheel	973 (0.56%)	327 (0.02%)	0.3	44 (3%)
Lithium-ion battery	5,324 (3.05%)	10,300 (0.6%)	2	770 (46%)
Lead-acid battery	76 (0.04%)	84 (0.01%)	1	89 (5%)
Sodium-based battery	208 (0.12%)	1,265 (0.08%)	6	69 (4%)
Flow battery	64 (0.04%)	177 (0.01%)	3	73 (4%)
Electro-chemical capacitor	32 (0.02%)	21 (0%)	0.7	27 (2%)
Nickel-based battery	89 (0.05%)	36 (0%)	0.4	10 (0.6%)
Thermal heat	2,487 (1.4%)	15,284 (0.9%)	6.1	81 (5%)
Latent heat	69 (0.04%)	461 (0.03%)	6.7	125 (7%)
Hydrogen storage	17 (0.01%)	100 (0.01%)	6	9 (0.5%)
Total	**174,376**	**1,655,763**	**9**	**1672**

Fig. 7.8 Operational energy storage power and storage capacity (2021, GWh). *Data Source* DOE Global Energy Storage Database, 2023

of renewable energy into the country's grid. Overall, the development of PHS continues to be driven by the increasing need for energy storage to support the integration of renewable energy into the grid and to provide grid stability and reliability.

In 2021, there were approximately 1672 operational energy storage projects. Although PHS projects accounted for only 22% of the total energy storage projects, they remained the most widely used form of utility-scale energy storage worldwide. The global installed power capacity of PHS was more than 164 GW, representing a dominant share of more than 94% of the global energy power capacity (174 GW). Together with other mechanical energy storage technologies, such as flywheel and compressed air energy storage, mechanical energy storage has an even larger share of 98%.

Most energy storage systems in operation today are used to provide daily balancing. Pumped hydropower storage is one of the energy storage technologies that has the longest energy storage duration among all operational energy storage facilities. The average power capacity (GW) and average storage capacity (GWh) of all operational pumped hydro projects were 164.4 GW and 1594 GWh, respectively, which account for an average storage duration of 9.7 h, which is longer than the average storage duration of 9 h. Considering PHS's dominant market position, its longer than average storage duration reflects the reality that most other energy storage technologies have shorter storage durations.

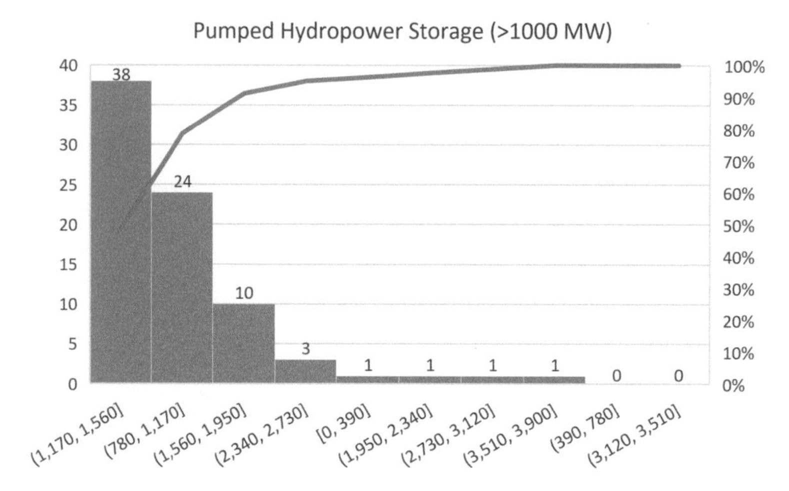

Fig. 7.9 Large pumped hydropower plants. *Data Source* Wikipedia, 2023

In 2022, the global PHS capacity reached 181 GW, including a 17 GW annual addition, which accounted for a 10.4% increase from the previous year's total capacity and almost quadrupled the amount added in 2021. There are 79 PHS plants in the world with an installed capacity of over 1 GW each and a total installed capacity of 107 GW. The majority (48%) of these large PHSs have an installed capacity in the range of 1.17–1.56 GW, 30% of them in the range of 0.78–1.17 GW, and 17% of them larger than 1.56 GW (Fig. 7.9).

By 2023, PHS remains the dominant technology for large-scale energy storage. Countries with significant pumped storage capacity included China, the U.S., Japan, and several European nations. However, other forms of energy storage, such as batteries, are becoming increasingly competitive, especially for shorter duration and smaller scale applications.

2. Battery Technologies

Lithium-ion (Li-ion) batteries have experienced rapid growth and adoption since their commercialization in the 1990s and have become the dominant technology in portable electronics, electric vehicles, and grid-scale energy storage. The installed capacity of Li-ion batteries in the world grew from less than 1 GWh in 2000 to over 10.3 GWh in 2021, with most of this growth occurring in the past decade. This growth has been driven by declining costs, improvements in energy density and cycle life, and policy support for energy storage deployment.

The first battery to store power, the voltaic pile, was invented by Alessandro Volta, an Italian chemist, in 1800. He created the device to produce a continuous electrical current for his scientific experiments. Stanley Whittingham, an English chemist at Exxon, invented the first lithium battery in the 1970s because of decades of research on the development of high-energy batteries with a longer cycle life and lighter weight. The growth of Li-ion batteries since then has been significant. In the early years, Li-ion batteries were primarily used in small-scale applications, such as calculators and watches. However, as technology advanced and the demand for portable electronics increased, the use of Li-ion batteries became more widespread.

In the 1990s, the first commercial Li-ion batteries were introduced, and they quickly gained popularity in consumer electronics such as laptops and mobile phones. As the technology continued to improve, Li-ion batteries became more efficient and affordable, leading to their adoption in other applications, such as electric vehicles and grid-scale energy storage. The battery price dropped by more than 79% from $732 in 2013 to $151 in 2022, which accounted for an annual average reduction rate of 7.9% in the decade (see Fig. 7.10).

The drastic price drop of electrochemical batteries using various materials, such as lithium, lead, and carbon, drove the rapid growth of battery storage projects, which accounted for 61.6% of all energy storage projects in 2021. However, these projects were of much smaller scales and have a storage power capacity of only 5.8 GW, accounting for a share of 3.3% of the total storage power capacity.

Today, Li-ion batteries are the most common type of lithium-based battery and have the largest share (approximately 92%) in lithium-based batteries in terms of power capacity. Approximately 87% of Li-ion batteries manufactured and distributed in 2021 were used in electric vehicles (EVs). The remaining 13% was split between consumer electronics, stationary storage, and other applications.

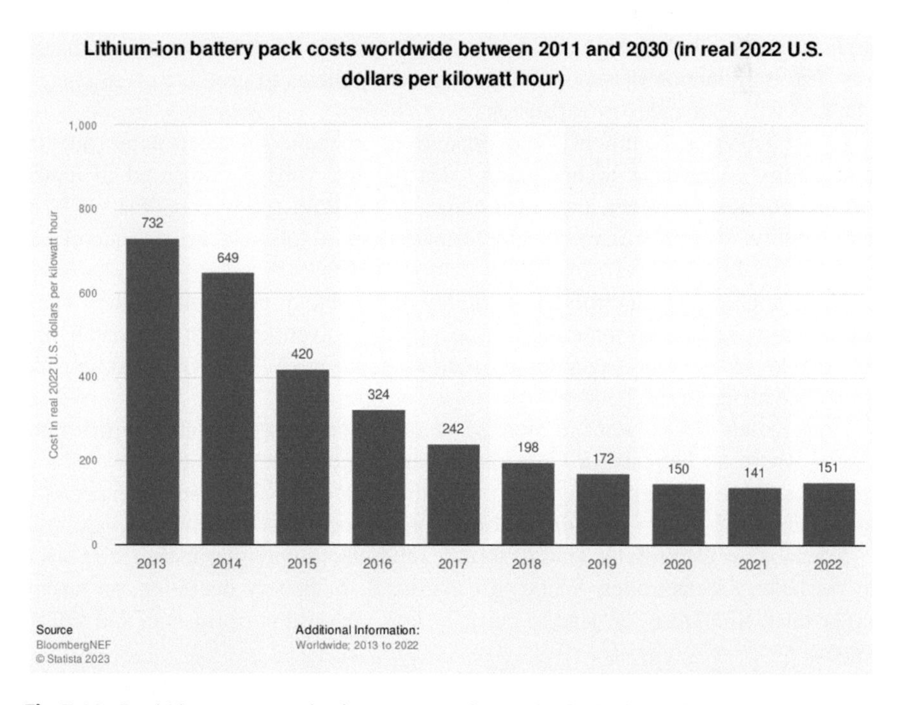

Fig. 7.10 Rapid battery cost reduction. *Image and Data Credit* Statistica, Bloomberg

Battery Type	Lithium-ion	Lead-Acid	Sodium-ion	Vanadium flow
Lifecycle	3,500	500 – 1,000	500 – 5,000	10,000
Energy Density (Wh/kg)	150 – 265	30 – 50	75 – 200	10 – 30
Cost ($/kWh)	151	100 – 300	47 – 77	500
Energy Efficiency	> 90%	70% – 80%	> 80%	> 60%
Safety	Organic electrolyte flammable, unsafe	Aqueous electrolyte unflammable but toxic	Aqueous electrolyte relatively safe	Aqueous electrolyte unflammable but toxic
Material	Scarce	Abundant	Abundant	Abundant

Fig. 7.11 Batteries by type, lifespan, density, efficiency, and safety

Grid-scale battery storage technology is typically employed for subhourly, hourly, and daily balancing. The U.S., China and Europe were the top three countries on the market, each accounting for gigawatt-scale additions. The estimated global energy storage capacity for batteries, with Li-ion batteries being the most dominant, was approximately 4.14 GW or 220 GWh, accounting for approximately 2.2% of the global energy storage capacity.

While the share of Li-ion batteries used for stationary storage is relatively small, the deployment of such systems is expected to grow rapidly in the coming years due to the increasing penetration of renewable energy sources and the need for grid stability. The demand for electric vehicles and renewable energy storage is the main driver of its growth, as Li-ion batteries are the preferred choice for these applications due to their high energy density, long cycle life, and low self-discharge rate. The introduction of supportive government policies to promote clean energy adoption will spur industry expansion.

Li-ion batteries commonly use organic or nonaqueous electrolytes due to their ability to provide higher energy density and voltage compared to aqueous electrolytes. However, they also come with certain safety concerns, such as flammability, thermal runaway, and formation of a solid-electrolyte interphase layer (Fig. 7.11).

Many nonaqueous electrolytes are highly flammable or even combustible. They consist of volatile and sometimes toxic organic solvents, which can catch fire or explode under certain conditions, such as exposure to heat, overcharging, or physical damage to the battery.

Nonaqueous electrolytes are more prone to a phenomenon called "thermal runaway," where an increase in temperature can trigger a self-accelerating reaction leading to battery overheating, fire, or even explosion. This can be especially dangerous if the battery is damaged or if it undergoes mechanical stress.

Nonaqueous electrolytes tend to form a solid-electrolyte interphase (SEI) layer on the battery's electrodes. While SEI is crucial for battery operation, an unstable or thick SEI layer can lead to capacity loss, reduced performance, and safety issues.

Nonaqueous electrolytes are sensitive to moisture and other contaminants. Even lesser amounts of water or impurities can degrade the performance and safety of the battery by promoting side reactions or causing short circuits.

Nonaqueous electrolytes may have a narrower operating temperature range compared to aqueous electrolytes. Extreme temperatures, whether too high or too low, can affect battery performance and safety.

Nonaqueous electrolytes can contribute to the formation of lithium metal plating and dendrites, which can grow on the battery's electrodes over time and potentially cause internal short circuits, leading to safety hazards.

Lead-acid, sodium-ion, and vanadium flow batteries use aqueous electrolytes, which avoid safety concerns, such as flammability, thermal runaway, and the formation of a solid-electrolyte interphase layer. However, these batteries also come with lower energy densities and energy efficiencies than lithium-ion batteries.

Aqueous electrolytes offer several potential advantages, such as improved safety due to their nonflammable nature compared to some nonaqueous electrolytes used in lithium-ion batteries. Additionally, aqueous electrolytes can be more environmentally friendly and cost-effective.

However, there are challenges associated with using aqueous electrolytes in sodium-ion batteries, including issues related to the wider electrochemical stability window of water-based electrolytes and the potential for limited cycling performance. Researchers continue to explore and optimize the use of aqueous electrolytes for sodium-ion batteries to overcome these challenges and improve overall battery performance.

3. Other Energy Storage Technologies

Other types of energy storage technologies, which include thermal energy storage, compressed air energy storage, flywheels, superconducting magnetic energy storage, and hydrogen storage, are still in research and development and are experiencing notable growth and interest due to increasing global demand for renewable energy integration, grid stability, and decarbonization efforts.

Thermal energy storage systems, including sensible (hot) and latent (cold) heat storage, involve materials such as molten salt in concentrated solar power generation and/or space- and water-heating applications or ice in cooling applications. Thermal energy storage technologies have also seen significant growth in recent years.

The global installed power capacity of thermal energy storage technologies was estimated to be approximately 2.5 GW, accounting for 1.4% of the global energy storage power capacity in 2021, and the global installed energy storage capacity of these systems grew from 1 GWh in 2000 to over 15 GWh in 2021, with most of this growth occurring in the U.S., Europe, and China.

Thermal energy storage technologies have shown increased applications and performance in CSP, space and water heating, and industrial and agricultural processing. Molten salt thermal storage had a higher average thermal energy duration of 7 h than that (6 h) of other types of thermal energy technologies.

Compressed air energy storage (CAES), another member of the mechanical energy storage group, represents an exception in terms of energy storage duration. Its average storage duration of more than 46 h is almost four times longer than that of PS. However, CAES has only a small market share of 0.4% or a small power capacity of 0.7% of the global energy storage capacity.

Challenges

A. Net Zero Requirement Challenge

According to the IEA's suggestion, by 2050, when 94% of power comes from renewable energy sources, the U.S. needs approximately 6 terawatt hours (TWh) of energy storage capacity, i.e., 930 GW for 6.5 h, to fully cover the power demand in the U.S. Considering the current energy storage capacity of 376 GWh in the U.S., the country needs to expand its energy storage capacity at an average growth rate of 55% per year. According to Bloomberg, an estimated 387 GW or 1143 GWh of new energy storage capacity will be needed globally from 2022 to 2030.

B. Technological Challenges

Although energy storage technologies for power applications have achieved various levels of technical and economic maturity in the marketplace, there are still many challenges. For grid storage, efficiency loss is at the top of the list of challenges. The roundtrip efficiencies of energy storage range from under 30% to over 90%. Efficiency losses represent a tradeoff between the increased cost of electricity cycled through storage and the increased value of greater dispatchability and other services to the grid.

The capital cost of many grid storage technologies is also exceedingly high relative to conventional alternatives, such as gas-fired power plants, which can be constructed quickly and are perceived as a low-risk investment by both regulated utilities and independent power producers. The existing market structures in the electric sector may also undervalue the many services that electricity storage can provide.

For transportation energy storage, the current primary challenges are the limited availability and excessive costs of both battery-electric and hydrogen-fueled vehicles. Additional challenges are new infrastructure requirements, particularly for hydrogen, which requires new distribution and fueling infrastructure. Battery electric vehicles are limited by range and charging times, especially when compared to conventional gasoline vehicles. Rechargeable batteries, also called secondary batteries, include lithium ions, lead acid, nickel cadmium, nickel metal hydride, sodium sulfur (NGK), sodium-ion, and molten metal batteries. Among these batteries, Li-ion and lead-acid batteries are the most popular and make up over 81% of new energy storage installations in the world. Most of these installations are for short duration applications (ranging from minutes to hours).

1. Pumped Hydropower Storage Challenges

Pumped hydropower storage (PHS) is one of the most established and widely used methods of energy storage. However, its adoption is limited due to various challenges.

Geographical and site availability poses significant hurdles for PHS facilities. Specific geographic conditions are needed, with suitable sites featuring two reservoirs at different elevations, preferably with large height differences. Identifying and securing appropriate locations can be challenging, especially in densely populated or environmentally sensitive areas. The competition for land use in these specific locations can be fierce, and not every region has access to such suitable sites.

High capital costs present another major challenge for PHS projects. Significant upfront investments are necessary for construction, equipment, and land acquisition. The capital costs can range between $500 million and $1.25 billion for a typical 500 MW facility, making it difficult for new projects to enter the market. Additionally, in regions where shorter-term storage needs are the focus, other energy storage options, such as Li-ion batteries, might be more cost-effective.

Environmental impact and misconceptions about it are also significant challenges for PHS expansion. Building PHS facilities may involve land use changes, alteration of waterways, and potential habitat disruption. These environmental concerns can lead to opposition from local communities or regulatory delays during the project development phase. Despite its crucial role in integrating renewables, PHS is often excluded from renewable energy targets and green incentives by governments.

PHS development timelines are lengthy, taking approximately seven years or more, mainly due to the rigorous licensing process and construction requirements. These delays hinder the ability to meet short-term grid balancing needs and keep pace with the rapid deployment of renewable energy.

Revenue calculation for PHS projects can be uncertain due to changing load and supply mixes, smaller price spreads between on- and off-peak periods, and the absence of long-term price signals for capital-intensive projects in wholesale markets.

While PHS has historically dominated the energy storage market, it now faces increasing *competition from other storage types*. The capital costs of utility-scale PV + battery storage, utility-scale battery storage, and commercial battery storage have become lower than those of PHS, and this trend is expected to continue.

Integrating PHS into the grid requires complex coordination with power generation sources, markets, and transmission systems, especially when grid infrastructure needs upgrades or modifications to accommodate bidirectional power flow during pumping and generation cycles. Significant upgrades to the power transmission system may be necessary to accommodate substantial amounts of variable renewable energy, which can be expensive and time-consuming.

Although PHS is known for its relatively high efficiency in energy storage and conversion, *some energy is inevitably lost* during the pumping and generation processes. The efficiency of the overall system can be affected by factors such as water evaporation and frictional losses in pipes and turbines.

2. Li-ion Battery Challenges

Although Li-ion batteries have become widely used in many applications, there are still some challenges that must be addressed to improve their performance and safety.

Limited Resource Availability. Lithium is a relatively rare metal that is found in limited quantities in the Earth's crust. Most of the world's lithium reserves are concentrated in a few countries, including Chile, Argentina, Australia, and China. The extraction of lithium can also be environmentally damaging, as it often requires substantial amounts of water and can lead to soil and water pollution.

While demand is expected to rise sharply, the availability of some raw materials used in Li-ion batteries, such as cobalt and lithium, is limited. For example, the demand for critical minerals, including lithium, cobalt, and nickel, could increase by up to six times by 2040. More than 60% of the world's known lithium reserves are in South America's lithium triangle, and approximately 70% of the world's cobalt supply comes from the Democratic Republic of Congo. This significantly surging demand could outstrip the available supply, leading to potential supply chain constraints and disruptions and increased costs.

The cost of Li-ion batteries has declined by approximately 88% from $1100/kWh in 2010 to approximately $152/kWh in 2022 but still accounts for approximately 40–50% of the total cost of an electric vehicle. Further cost reductions are needed to make renewable energy systems more competitive with fossil fuels. However, the resource availability of cobalt, lithium, nickel, and copper, as well as the complex manufacturing processes involved, could prevent the further cost reduction of Li-ion batteries.

Recycling. According to a report by the International Energy Agency, only approximately 5% of Li-ion batteries are currently recycled, leading to a significant loss of valuable materials. Increasing the recycling rate of Li-ion batteries is crucial to recover these materials and reduce the environmental impact of battery manufacturing and disposal.

Safety Issues. According to a report by the National Renewable Energy Laboratory, the risk of thermal runaway in Li-ion batteries is low, but it can have severe consequences, especially for air transportation.

Energy density. According to a report by the U.S. Department of Energy, the energy density of Li-ion batteries has increased from 100 Wh/kg in 1990 to over 250 Wh/kg in 2020. However, further improvements are needed to meet the energy storage requirements of certain applications, such as long-range electric vehicles and grid-scale energy storage systems.

Charging infrastructure. According to the International Energy Agency, the number of electric vehicle chargers worldwide is expected to reach 40 million by 2030, up from approximately 7 million in. 2019. However, significant investment is still needed to ensure the availability of charging infrastructure to support the widespread adoption of electric vehicles and energy storage systems.

3. Superconducting Magnetic Energy Storage

SMES is an emerging technology and has not yet gained significant market traction. The major limitation of SMES is its high cost, which is mainly due to the expensive materials used in the superconducting coil and the need for cryogenic cooling. Nonetheless, research is being conducted to improve the technology and reduce costs, making it a promising energy storage solution for the future.

4. Hydrogen and Fuel Cell Challenges

Hydrogen has been considered a potential energy storage medium for many years due to its high energy density and its potential to be produced from renewable energy sources. However, the use of hydrogen as an energy storage solution remains limited because of major challenges in the decarbonization of hydrogen generation, the infrastructure of hydrogen generation, storage, transportation, and fuel cell-based hydrogen utilization (Fig. 7.12).

Green hydrogen generation is a primary challenge. Hydrogen is not readily available in nature and must be generated using various processes and technologies. Currently, approximately 96% of hydrogen production relies on using fossil fuels as feedstock or fuel, mostly through steam reformation of natural gas. This

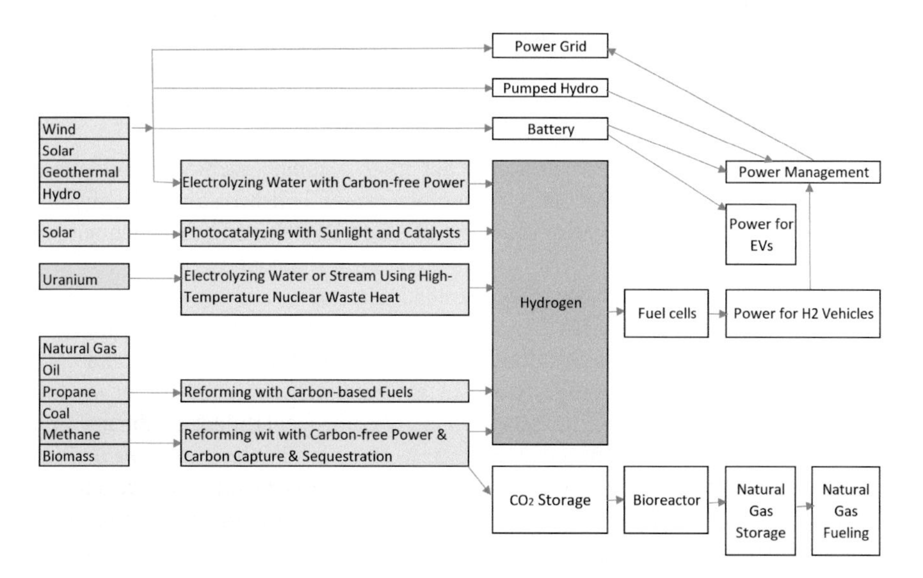

Fig. 7.12 Pathways of hydrogen generation: green, blue, and gray

reliance has significant greenhouse gas emissions, which defeats our goal of decarbonizing energy production and consumption. Green hydrogen generation that uses surplus wind and solar power to split water (a process called electrolysis) makes the hydrogen generation and energy storage solution more sustainable. However, green hydrogen was estimated to account for approximately 0.1% of the total hydrogen production worldwide.

Hydrogen infrastructure is a major barrier. Because hydrogen is not found naturally in its pure form and must be produced through steam methane reforming or electrolysis, it needs specialized and costly infrastructure for its generation, transportation, and storage before it can be used as a fuel. These infrastructures and technologies include hydrogen generation facilities, storage tanks, pipelines or transport vessels, dispensing stations, and storage technologies such as hydrogen compression and liquefication. In addition, as a highly flammable gas, hydrogen requires special handling and safety measures to prevent accidents.

Hydrogen storage refers to technologies for storing hydrogen in different forms as a fuel source for later use. Hydrogen storage technologies include compressed hydrogen storage, liquid hydrogen storage, solid-state metal hydrate, carbon nanotube storage, and chemical hydrogen storage.

Hydrogen storage technologies involve various mechanical requirements, including low temperatures, high pressures, or the use of chemical compounds that release hydrogen only when necessary. It is most widely used in manufacturing sites, especially in the synthesis of ammonia, to generate power or drive electric vehicles.

However, hydrogen storage has significant challenges associated with storing hydrogen with compression or liquefaction. Storing hydrogen as 350-bar compressed gas requires high-pressure storage tanks and has a round-trip efficiency of just 47%. Storing hydrogen as a liquid requires a cryogenic (extremely low) temperature of − 253 °C at 1 bar and has a higher round-trip efficiency of 70%, meaning that 30% of hydrogen is lost. Both hydrogen storage technologies have lower efficiencies than batteries, i.e., 70–90%.

Compressed hydrogen gas storage is the most common method of hydrogen storage for fuel cell vehicles. This technology stores hydrogen in high-pressure tanks, which compress the gas to pressures of up to 10,000 psi (pounds per square inch).

Liquid hydrogen storage technology stores hydrogen as a cryogenic (i.e., deep-freezing) liquid, which requires extremely low temperatures (− 253 °C/− 423 °F) to liquify hydrogen gas. The resulting liquid hydrogen is stored in insulated containers. Liquid hydrogen storage is one of the most efficient methods of storing hydrogen because it has an extremely high energy density. This allows the use of a relatively small amount of space to store a large amount of hydrogen. However, liquid hydrogen storage also requires careful handling and storage because extremely low temperatures can pose safety risks and can lead to evaporative losses.

Liquid hydrogen storage is used in a variety of applications, including in the aerospace industry for rocket propulsion and in fuel cell vehicles. It is also used in some industrial processes and research applications. However, due to the challenges involved in handling and storing liquid hydrogen, other methods of hydrogen storage, such as compressed gas and solid-state storage, are also being developed.

Metal hydride technology stores hydrogen in solid-state metal hydrides, which absorb hydrogen and release it when heated. Metal hydride storage is a safe and compact way to store hydrogen, but its storage capacity is limited.

Carbon nanotube storage technology, which is still in the experimental stage, stores hydrogen in carbon nanotubes, which have a large surface area and can absorb hydrogen through attraction between atoms, molecules, and surfaces, as well as other intermolecular forces.

Chemical hydrogen storage technology stores hydrogen in chemical compounds, such as ammonia or methanol, which release hydrogen when heated or reacted with a catalyst. This technology has the potential for high-density hydrogen storage but requires additional processing steps to release hydrogen.

Each of these storage methods has its own advantages and limitations, and the choice of storage method depends on the specific application and requirements. Important applications of hydrogen storage systems are fuel in the transportation sector, power supply in the industrial sector, and heating in the residential sector.

Hydrogen fuel cell. Hydrogen applications, such as transportation, power generation, industrial processes, energy storage, and grid stabilization for stationary power generation, require specialized and costly devices such as hydrogen fuel cells. There are significant challenges to making hydrogen fuel cells a viable alternative to traditional fossil fuels. Currently, the cost of producing hydrogen fuel cells is approximately $80–100 per kilowatt (kW). To be competitive with traditional fossil fuels, the cost of fuel cells must be reduced to less than $50 per kW. In comparison, the cost of producing power from natural gas power plants is approximately $20–30 per kW.

Metal hydrides are a class of materials that can be used to store hydrogen for fuel. These materials absorb and release hydrogen through reversible chemical reactions, making them a promising option for hydrogen storage.

To utilize the hydrogen stored in metal hydrides as fuel, the metal hydrides must first be exposed to a suitable temperature and pressure to release the hydrogen. This process is known as desorption. The released hydrogen can then be used in a fuel cell or burned in a combustion engine to produce energy.

The process of desorption can be optimized by controlling factors such as temperature, pressure, and the composition of the MH. Different metal hydrides have different desorption characteristics, and some may require higher temperatures or pressures to release hydrogen.

In addition to optimizing desorption, it is also important to ensure that the metal hydrides are stable and durable over multiple cycles of hydrogen absorption and desorption. Research is ongoing to develop metal hydrides that are both efficient in storing and releasing hydrogen and are also durable and stable.

Overall, utilizing the hydrogen stored in metal hydrides as fuel requires careful control of desorption conditions and the selection of appropriate materials with good stability and durability. With further research and development, metal hydrides could become a vital component of future hydrogen storage systems.

5. Long-Duration Energy Storage Challenges

Long-duration energy storage refers to energy storage systems that can provide several hours to a few days of energy storage. Long-duration energy storage can be useful both in off-grid applications and in on-grid applications.

In recent years, there has been a shift toward using cleaner and more sustainable energy storage technologies for long-duration energy storage, such as PHS, compressed air energy storage, flow batteries, hydrogen fuel cells, and thermal energy storage. However, there are multiple challenges that need to be addressed to fully realize its potential in meeting the needs of both off-grid and on-grid markets.

Environmental Impact. Some long-duration energy storage technologies, such as pumped hydro and compressed air energy storage, have significant environmental impacts, particularly in terms of land use, water consumption, GHG emissions, and soil contamination.

High Cost. Long-duration energy storage technologies require significant upfront investments and development, manufacture, and deployment, which make them cost-prohibitive for some applications.

Performance Issues. Some energy storage technologies, such as Li-ion batteries, have efficiency issues, such as low density, low energy efficiency, and long charging time, which makes them uncompetitive in their service as long-duration energy storage, meaning that they need more response time, and some of the stored energy is lost during the storage and discharge process.

Scale. Long-duration energy storage systems face the contradiction between the scale needed for economic reasons and the land needed to accommodate such large energy storage systems. This contradiction often makes it impossible to deploy them in densely populated urban areas and to upscale them to rapidly reduce investment costs.

Safety. Long-duration energy storage systems can pose safety risks, particularly if they use flammable or hazardous materials.

Technical Challenges. Some long-duration energy storage technologies, such as molten salt energy storage and phase change material energy storage, face technical challenges associated with their design and operation, including issues related to heat transfer, thermal losses, and material durability. These challenges can impact the efficiency and reliability of energy storage systems.

For example, molten salts can be corrosive to certain materials, which can cause leakage-related operational issues, and solidify at low temperatures, which can cause flow and heat transfer problems. Phase change materials used in PCM energy storage systems can have a limited thermal conductivity, limiting the rate of heat transfer and experiencing degradation over time due to repeated heating and cooling cycles impacting the durability of the system.

Solutions

A. Expanding Competitive Energy Storage Markets

Expanding the deployment of energy storage technologies offers significant environmental benefits. In fact, the global energy storage market could attract more than $100 billion by 2024. The main reason for this growth is the replacement of diesel and fossil fuel generation with reliable renewable energy, which helps eliminate greenhouse gas emissions. This shift could reduce emissions by one million tons of CO_2 per year for every 1 GW of installed systems.

The DOE Sandia Laboratory showed through its recent research data that the installed costs of most energy storage technologies are already competitive for various energy storage systems and applications. While PHS and compressed air storage technologies are limited to large-scale energy storage demand, electrochemical energy storage technologies, such as lithium-ion, sodium, nickel, and zinc batteries, are already cost-effective for energy storage at different scales, from wholesale and utility to business and residential systems (Fig. 7.13).

B. Economies of Scale

Research found that there is no significant difference between the predevelopment, interconnection and maintenance costs and soft costs such as personnel training for a pilot utility-scale energy storage system and those costs for a 10 MW system. Therefore, the return on investment (profit minus cost) for a larger project is much better than that for a small project. Therefore, to achieve economies of scale, the installation of utility-scale energy storage needs to be large.

The National Renewable Energy Laboratory (NREL) found that except for PHS, energy storage technologies such as utility-scale PV-Plus batteries, utility-scale battery storage, commercial battery storage, and residential battery storage will all significantly reduce their capital costs in 2030, 2040, and 2050 (Fig. 7.14).

C. Government Support and Guidance

It is important to realize that promotional government legislation, policy, and funding can help significantly improve the economics of expanding energy storage technologies. The EU has been promoting energy storage through various directives and regulations. The Clean Energy for All Europeans package includes provisions to support energy storage and flexibility resources. The Horizon 2020 program also provided substantial funding for research and innovation in energy storage. The Chinese government implemented policies to promote energy storage, especially as a part of its commitment to increase the share of nonfossil fuels in primary energy consumption and to peak carbon dioxide emissions. Various provincial governments in China also had local programs and policies to support energy storage.

Technology	Wholesale (100 MW)	Utility (10 MW)	Microgrid (1 MW)	Business (0.1 MW)	Residential (0.01 MW)
Pumped Hydro Storage (PHS)	$ 1,633.2				
Compressed Air Energy Storage (CAES)	$ 1,614.3				
Sodium (Na) / 6h	$ 378.5	$ 400.7	$ 439.9		
Zinc (Zn) / 4h	$ 252.1	$ 268.0	$ 310.7	$ 366.1	
Long-Duration Flywheel (FWLD)		$ 676.0	$ 753.3		
Short-Duration Flywheel (FWSD)		$ 880.0	$ 1,146.5	$ 1,250.0	
Vanadium Flow Battery (FBV) / 4h	$ 409.9	$ 424.2	$ 520.1	$ 617.3	
Vanadium Flow Battery (FBV) / 6h	$ 345.5	$ 379.1	$ 432.1	$ 522.5	
Vanadium Flow Battery (FBV) / 8h	$ 304.6	$ 343.1	$ 402.4	$ 487.2	
Zinc Bromide Flow Battery (FBZnBr) / 3h					$ 746.5
Zinc Bromide Flow Battery (FBZnBr) / 4h	$ 472.2	$ 506.0	$ 555.3	$ 630.3	
Zinc Bromide Flow Battery (FBZnBr) / 6h	$ 417.2	$ 429.5	$ 452.0	$ 493.9	
Zinc Bromide Flow Battery (FBZnBr) / 8h	$ 413.5	$ 425.7	$ 447.3	$ 485.9	
Iron Flow Battery (FBFe) / 4h	$ 404.5	$ 416.4	$ 443.8	$ 505.1	
Nickel (Ni) Battery / 2h			$ 339.6	$ 466.4	$ 835.2
Nickel (Ni) Battery / 3h			$ 320.5	$ 434.4	$ 741.6
Nickel (Ni) Battery / 4h			$ 311.0	$ 418.5	$ 694.8
Lithium-Ion Energy Battery / 2h	$ 451.8	$ 474.6	$ 558.9	$ 719.4	$ 1,057.8
Lithium-Ion Energy Battery / 3h	$ 429.3	$ 449.9	$ 535.5	$ 679.5	$ 930.9
Lithium-Ion Energy Battery / 4h	$ 420.1	$ 440.6	$ 523.8	$ 659.6	$ 867.4
Lithium-Ion Energy Battery / 5h	$ 410.5	$ 435.0	$ 516.8	$ 647.6	$ 829.3
Lithium-Ion Energy Battery / 6h	$ 406.8	$ 431.2	$ 512.1	$ 639.6	$ 803.9
Lithium-Ion Energy Battery / 7h	$ 404.2	$ 428.6	$ 508.8	$ 633.9	$ 785.8
Lithium-Ion Energy Battery / 8h	$ 402.2	$ 426.6	$ 506.3	$ 629.6	$ 772.2
Lithium-Ion Power Battery / 0.5h	$ 808.5	$ 849.8	$ 1,007.3	$ 1,398.9	
Lithium-Ion Power Battery / 1h	$ 466.3	$ 485.5	$ 592.3	$ 755.2	
Lead (Pb) Battery / 2h			$ 445.8	$ 550.4	$ 810.6
Lead (Pb) Battery / 3h			$ 375.5	$ 461.8	$ 633.2
Lead (Pb) Battery / 4h			$ 338.4	$ 415.4	$ 542.3
Lead Carbon (PbC) Battery / 2h			$ 720.9	$ 877.8	$ 1,201.6
Lead Carbon (PbC) Battery / 3h			$ 598.1	$ 738.4	$ 973.8
Lead Carbon (PbC) Battery / 4h			$ 547.5	$ 664.4	$ 855.0

Fig. 7.13 Competitive energy storage markets—installed costs. *Data Source* DOE Sandia, 2021

The Australian Renewable Energy Agency (ARENA) and the Clean Energy Finance Corporation (CEFC) provided financial support for energy storage projects, especially those integrated with renewable energy. The Indian government, through schemes such as the National Electric Mobility Mission Plan (NEMMP) and the Faster Adoption and Manufacturing of Hybrid and Electric Vehicles (FAME), aimed to promote energy storage. South Korea has been actively promoting energy storage, with ambitious targets and financial incentives for energy storage systems. Other countries also had similar programs and policies to promote energy storage, and international agencies such as the International Renewable Energy Agency (IRENA) and the World Bank also provided support.

To support the research and development of energy storage, the U.S. government provided investment tax credits or production tax credits. To strategically

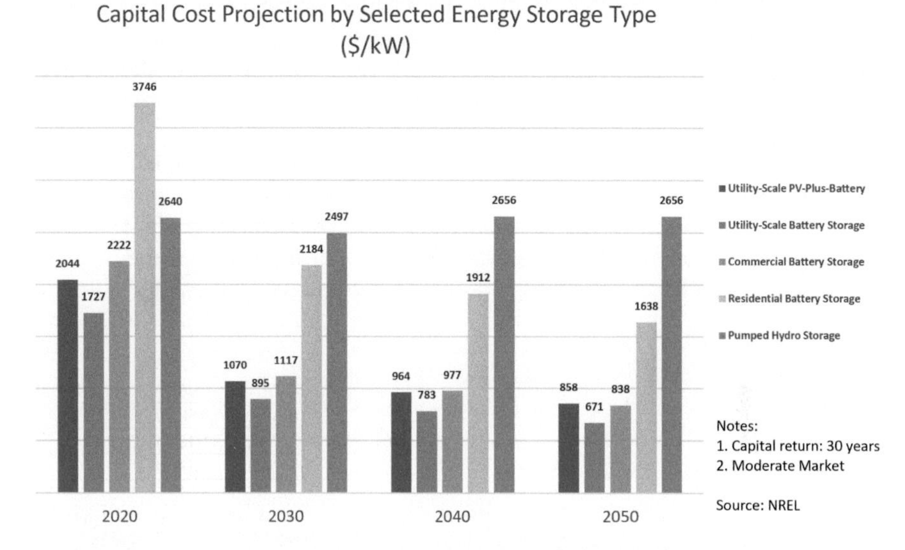

Fig. 7.14 Energy storage capital cost projection between 2020 and 2050

drive the growth of energy storage capacity in the U.S., DOE established the Energy Storage Grand Challenges (ESGC) in 2020, which set potential price targets for energy storage in the U.S. (Fig. 7.15).

D. Technological Solutions

1. Pumped Hydropower Storage Solutions

Recognizing the Importance of PHS. Expanding investments in PHS and its power and storage capacity offer substantial benefits. Increased PHS capacity can improve grid stability by efficiently managing fluctuations in renewable energy generation and electricity demand. As renewable energy adoption continues to grow, effective and reliable energy storage solutions such as PHS become even more critical.

Facilitating an Evolving Grid	Critical Services	Interdependent Network Infrastructure	Electric Mobility	Remote Communities	Facility Flexibility, Efficiency, and Value Enhancement
• \$0.03-\$0.05/kWh levelized cost of storage	• \$77/kW-yr for reliability applications • \$13912/kW-yr to offset backup generators	• \$77/kW-yr storage capex	• \$80/kWh manufacture cost for battery pack • \$104/kW-yr storage capex	• \$65/MWh delivered energy	• \$85/kWh, \$52/kW-yr for commercial and residential buildings • \$20-\$52/kW-yr for energy intensive facilities

Fig. 7.15 Potential price targets for energy storage in the U.S. *Data Source* U.S. Department of Energy

PHS enables better integration of renewable energy sources, reducing reliance on fossil fuels and contributing to greenhouse gas emissions reduction. A robust PHS infrastructure enhances energy security by providing backup power during peak demand or emergencies. Expanding PHS projects can create job opportunities and stimulate economic growth in the energy sector.

Along with its proven technology and benefits, the current small share of PHS installations in total hydropower capacity suggests numerous opportunities for further development and expansion. PHS has significant potential for large-scale energy storage, grid balancing, and integration of renewable energy sources, making it an asset for transitioning to a sustainable and resilient energy system.

Case Study

The RE100 Group of the Australian National University uses geographic information system (GIS) algorithms with defined search criteria to identify potentially feasible PHS sites (Fig. 7.16).

Fig. 7.16 PHS Atlas. *Image Credit* re100.anu.edu.au

This project is divided into the following types of sites: greenfield, bluefield, brownfield, and ocean sites. In 2023, the team has already completed the survey for the first two types of sites, having identified approximately 616,000 potential greenfield PHS sites with a storage potential of approximately 23 million GWh and 530,000 bluefield PHS sites with a storage potential of approximately 22 million GWh.

The potential capacity of each of these two types is approximately 100 times greater than required to support a 100% global renewable power system. Each site has a pair of upper and lower reservoirs plus a hypothetical tunnel route between the reservoirs. The group developed an online map for these identified sites along with data such as latitude, longitude, altitude, head, slope, water volume, water area, rock volume, dam wall length, water/rock ratio, energy storage potential and approximate relative cost. The survey reveals that almost all upper reservoirs of the identified sites are off-river, and none of them intrude into protected or urban areas. This means that at least 50% of each identified site has a closed cycle and is more environmentally friendly than regular hydropower generation.

The untapped potential of PHS lies in its unique ability to store excess energy during times of low demand and release it when demand is high. This flexibility allows PHS to act as a valuable backup to intermittent renewable energy sources such as solar and wind, providing stability and reliability to the grid. Moreover, PHS can play a crucial role in addressing energy supply–demand imbalances and supporting the integration of more renewable energy into the grid.

Addressing PHS challenges. Pumped hydropower storage (PHS) remains a valuable and reliable method of energy storage. However, it is essential to address the challenges discussed earlier, such as site availability, high capital costs, and environmental considerations, through developing and implementing innovations in policy making, market incentives, reviews, licensing, and public awareness.

Site Selection. To reduce siting complications, environmental impact assessments for new PHS projects need to conduct thorough studies to identify potential impacts and develop effective mitigation measures. Sites should avoid overly sensitive or scenic locations, and consideration can be given to reusing disused or abandoned industrial areas, such as steel mills, refineries, landfills, and mines.

New PHS projects should follow a "closed loop" PHS approach, utilizing two reservoirs at different elevations that are physically separated from existing river systems. This design minimizes significant riverine impacts associated with PHS deployment. Advances in geospatial analysis and machine learning algorithms can aid in identifying suitable locations for new PHS projects.

Advances in geospatial analysis and machine learning algorithms can also be used to identify suitable locations for new PHS projects. In Australia, for example, a machine learning algorithm was developed to identify potential PHS sites using a range of environmental, technical, and economic factors.

Further reducing PHS costs. Cost reduction efforts include technological innovation and economies of scale. For example, in China, the State Grid Corporation has developed a new type of PHS technology that uses seawater as the storage medium, which is expected to significantly reduce capital costs compared to traditional PHS.

Operational flexibility. New PHS projects are being designed to provide greater operational flexibility, including faster ramp rates[3] and the ability to provide ancillary services to the grid. For example, in the U.S., the Grid Modernization Initiative is funding the research and development of new PHS technologies that can provide fast-ramping, flexible energy storage.

[3] Ramp Rate refers to the rate of ramping up, rate of increasing or changing speed.

Regulatory and market frameworks. Policymakers, governments, and energy companies should consider the untapped potential of PHS and explore ways to incentivize its development and deployment. Governments can establish policies and market mechanisms to encourage investment in PHS projects, making them economically viable. Various countries have implemented pricing mechanisms and auctions to incentivize energy storage technologies, including PHS.

Governments can also develop methodologies to fairly assess the value of energy storage and capacity products across all energy storage technologies. Simplifying the licensing process for low-impact PHS projects can reduce red tape.

It is important to note that PHS is the most cost-effective long-duration large-scale energy storage technology. To help overcome the challenge of its high upfront costs and ensure that it can indeed compete with other energy storage technologies, governments and energy regulators can establish and implement policies and market mechanisms to encourage greater investment in PHS projects. For example, governments can pass short- and long-term legal targets or financial incentive programs for all energy storage technologies, including PHS, on a technology-neutral basis.

For example, in China, two prices are implemented, i.e., one storage price charged by grid companies at 70% of the power price and the other competitive market power price charged by storage power plants, which allows for energy storage companies to increase their profit by 5–10%. In the UK, capacity market auctions provide financial incentives for energy storage technologies, including PHS. In the U.S., the Federal Energy Regulatory Commission has issued rules that enable PHS projects to participate in wholesale electricity markets, providing additional revenue streams and making these projects more economically viable.

Governments can also develop a feasible methodology of assessment to mitigate the challenge of fairly assessing the value of energy storage and capacity products across all energy storage technologies. To reduce red tape in licensing and permitting, governments should continue to simplify the licensing process for low-impact PHS, such as off-channel, modular or closed-loop projects. The PHS industry can also select a standard type of pumped storage to help all decision-makers better understand and compare PHS technologies.

The International Forum on Pumped Storage Hydropower, for example, attempted to develop guidance and recommendations on how PSH can best support future power systems in the clean energy transition in the most sustainable way. It recommends the sustainability assessment of PSH projects using a multilevel approach, including system-level needs option assessment and project optimization, the Hydropower Sustainability Standard and other existing hydropower sustainability assessment tools.

At the same time, the organization also rejects capital expenditure (CAPEX) comparisons because these simplistic studies ignore the shorter cycle life and higher replacement and maintenance costs of other energy storage technologies.

For example, PSH can typically be effectively used for 80 years, while batteries normally only have a lifespan of approximately 10 years. On the other hand, PSH typically has a storage capacity of GWh suitable for large-scale utility grids, while batteries only have a storage capacity in MWh that can be used for microgrids.

In addition to these comparisons, the working group also provided a brief series of peer reviews of innovations for PSH. These innovative approaches for PSH covered three broad categories: furthering PSH potential (such as seawater PSH), retrofitting and upgrading PSH systems (such as utilizing abandoned mines), and developing hybrid systems (such as combined with thermal storage).

Policy and Markets Working Group's Policy Recommendations. The Policy and Markets Working Group of the International Forum on Pumped Storage Hydropower (IFPSH) put forward seven major recommendations for governments around the world to avert the risk of policymakers and grid operators falling back on fossil fuels to provide clean energy storage:

- Assess long-term storage needs now so that the most efficient options, which may take longer to build, are not lost.
- Ensure consistent, technology-neutral comparisons between energy storage and flexibility options.
- Remunerate providers of essential electricity grid, storage and flexibility services.
- Licensing and permitting should take advantage of internationally recognized sustainability tools.
- Ensure long-term revenue visibility with risk sharing to deliver the lowest overall cost to society.
- Assess and map pumped hydropower storage among potential existing hydropower assets and prospective sites.
- Support and incentivize pumped hydropower storage in green recovery programs and green finance mechanisms.

2. Battery Solutions

The challenges facing the battery are not insurmountable. Currently, many viable solutions are being researched and developed.

Regulatory and policy support is the cornerstone of many solutions. Implementing carbon pricing policies, for example, can reduce the relative cost of renewable energy technologies, including energy storage technologies, and accelerate the transition to a net-zero scenario by 2050. At the same time, documents, policies, and strategies issued by various international organizations and national governments to promote energy storage can also play a vital role in accelerating the development and deployment of energy storage technologies.

The U.S. Department of Energy developed a strategy to accelerate the development and deployment of energy storage technologies in the U.S. It outlines a comprehensive approach to achieve the Grand Challenge goal of developing and deploying energy storage systems that are five times more energy dense, five times cheaper, and able to cycle five times faster than current technologies by 2030.

The roadmap identifies various technical and nontechnical barriers that need to be addressed to achieve these goals, including advancing research and development, improving manufacturing and supply chain capabilities, addressing safety and environmental concerns, and improving market design and policy frameworks.

To overcome these barriers, the roadmap suggests a multipronged approach that includes increasing research and development funding, improving testing and validation procedures, establishing technical standards, and supporting workforce development initiatives.

Furthermore, the roadmap highlights the importance of partnerships between industry, academia, and government to advance energy storage technology development and deployment. The roadmap envisions a future where energy storage systems play a crucial role in ensuring a more reliable, resilient, and sustainable energy system, benefiting American households, businesses, and the economy.

Technological innovation is an important strategy in fuel cell solutions. Developing new materials and manufacturing processes can reduce the cost of Li-ion batteries from approximately $137/kWh to as low as $62/kWh by 2030, according to a report by Bloomberg New Energy Finance. Improved battery management systems could also extend the lifetime and performance of batteries.

Fostering sustainable and circular supply chains is key to the sustainable deployment of battery technology. Increasing the recycling rate of Li-ion batteries from 5 to 50% can recover up to 10 million tons of materials by 2030, according to a report by the International Energy Agency. A company has recycled lithium metal and Li-ion batteries since 1992 at its facility in British Columbia, Canada. In 2015, the company began operating the first U.S. recycling facility for Li-ion vehicle batteries in Lancaster, OH. Approximately 25 companies in North America and Europe recycle lithium batteries or plan to do so. Partnerships between automobile companies and battery recyclers have been made to supply the automobile industry with a source of battery materials. This can significantly reduce the demand for critical raw materials and minimize environmental impacts.

Collaborative efforts. Collaboration between industry stakeholders, policymakers, and researchers could help accelerate the development and adoption of new battery technologies. For example, the European Battery Alliance, a public–private partnership, aims to develop a sustainable and competitive battery value chain in Europe.

Scaling up infrastructure. Scaling up the infrastructure for charging electric vehicles and energy storage systems could require significant investment. For example, according to the International Energy Agency, an investment of approximately $300 billion is needed to develop a robust charging infrastructure network worldwide by 2030.

Implementing these solutions will require significant investment, collaboration, and policy support, but they have the potential to accelerate the transition to a net-zero scenario by 2050 and create a more sustainable and efficient energy system.

Addressing the safety issues of Li-ion batteries needs to adopt a two-step approach, finding short-term solutions and long-term solutions. Short-term solutions include research and development of better battery management systems and the implementation of rigorous safety standards. The long-term solutions include research and development of alternative advanced batteries.

Short-term solutions primarily revolve around optimizing existing technologies. This includes extensive research and development efforts aimed at creating advanced battery management systems. These systems would enable more efficient monitoring and control of battery parameters, such as temperature and charge state, thereby reducing the risk of battery failure and improving overall safety. In tandem with this, it is imperative to enforce stringent safety standards that all Li-ion batteries must adhere to. This ensures that any battery reaching the consumer market has undergone rigorous testing to confirm its safety and reliability.

Long-term solutions involve pivoting away from Li-ion technology altogether. This would entail research and development of alternative, advanced battery technologies that are inherently safer. Technologies such as solid-state batteries and graphene-based batteries, for example, hold promise due to their higher energy densities, faster charge times, and reduced risk of fire or explosion. By investing in these technologies, we can pave the way for safer, more efficient batteries in the future.

Alternative battery chemistries. Developing alternative battery chemistries, such as solid-state batteries, could increase the energy density of batteries by up to 50%, according to a report by the U.S. Department of Energy. This could help meet the energy storage requirements of certain applications, such as long-range electric vehicles and grid-scale energy storage systems.

(a) Solid-State Batteries

Solid-state batteries hold the potential to address many concerns of traditional Li-ion batteries, including improved energy density, safety, and lifespan. They use solid electrolytes instead of liquid electrolytes. The inherent nonflammability reduces the risk of thermal runaway, increasing the design flexibility and volumetric density. Solid-state batteries also maintain high conductivity at subzero temperatures down to − 20 °C (− 4 °F), overcoming a significant shortcoming of standard electric vehicle (EV) batteries.

While companies such as Ford, Lilium, Eviation, QuantumScape (in partnership with Volkswagen), Solid Power (backed by Ford and BMW), and Toyota are developing this technology, solid-state batteries remain largely in the research and development stage. The potential of this technology deemed the next significant advancement in EVs and portable electronics hinges on overcoming several technical and commercial challenges.

Current solid-state batteries struggle to match the performance of Li-ion batteries in terms of energy density, power output, and cycle life. The energy density of top-performing solid-state batteries is currently approximately 400 Wh/kg, while Li-ion batteries can achieve over 700 Wh/kg. Their inherently brittle nature makes them prone to cracks and fractures, potentially leading to reduced performance and shorter lifespan.

Manufacturing solid-state batteries is currently more expensive than manufacturing traditional Li-ion batteries, largely due to the higher material costs and underdeveloped manufacturing equipment. The cost of materials for solid-state batteries is estimated to be 2–3 times higher than that for Li-ion batteries. Scaling up production to meet demand is another significant challenge, with large-scale production potentially still 5–10 years away.

Building the necessary manufacturing infrastructure also requires significant investment and time, and the availability of new materials to replace the liquid electrolytes used in traditional Li-ion batteries could limit the production and commercialization of solid-state batteries. For instance, some solid-state electrolytes require rare or expensive materials, such as lithium-phosphorus oxynitride or lithium-garnet. The dominance of solid-state batteries over Li-ion batteries will depend on factors such as EV industry demand and initial costs.

To surmount the challenges of graphene batteries, continued research and development, investment in manufacturing infrastructure, and establishing regulatory frameworks that ensure safe and sustainable use of graphene-based products are vital.

Significant R&D efforts are already underway to address these issues, exploring more cost-effective methods for high-quality graphene production, developing a robust manufacturing infrastructure, and establishing production and characterization standards for graphene. Moreover, researchers are working on improving the stability and durability of graphene-based batteries and establishing regulations for their safe handling and disposal.

Despite many challenges, the unique properties and potential benefits of graphene batteries make them a promising contender for the future, playing a significant role in the transition to renewable energy and sustainable development. However, the transition from the research and development stage to large-scale production and commercialization remains a formidable challenge.

(b) Lithium-Sulfur Batteries

Lithium-sulfur (Li–S) batteries have higher energy densities and cost-effectiveness than lithium-ion batteries. They use sulfur as the cathode material, which can reduce the weight of batteries with greater capacity. Despite the potential to double current battery ranges, however, these batteries still need to address challenges related to cycle life and technological maturity because the expansion and contraction of sulfur during charging and discharging can cause degradation. They are still in the experimental stage and have not yet reached commercialization.

(c) Silicon Anode Lithium-Ion Batteries

Silicon anode lithium-ion batteries are redefining the capacity and range potential. By integrating silicon-based anodes, these batteries achieve a significant boost in energy density and performance due to their higher lithium capacity compared to traditional graphite anodes. However, silicon anodes also suffer from large volume changes during cycling, leading to capacity fade and mechanical stress. The challenges of managing silicon's expansion and reactivity remain to be mitigated in ongoing research and development.

(d) Lithium Iron Phosphate Batteries

Lithium iron phosphate (also known as LFP or LiFePO4) batteries, which first appeared in 1996, are replacing other battery technologies due to their technical advantages and extremely high level of thermal and chemical stability. These features provide better safety than Li-ion batteries made with other cathode materials. They are commonly used in electric vehicles and energy storage systems. LFP chemistry also offers a longer cycle life than other types of Li-ion batteries. The use of iron (Fe) is safer, cheaper, and more environmentally friendly than the use of cobalt and manganese. However, LFP batteries have lower energy density than NCM and NCA batteries. They also have lower voltage and power output, which might be a disadvantage in high-performance applications. Similar to other battery types, LFP batteries are produced and consumed globally but have a significant presence in China due to favorable safety characteristics and the absence of expensive metals. Some of the leading companies producing LFP batteries include BYD, CATL, and Tesla (in their China-made Model 3 and Model Y vehicles). LFP batteries have lower energy density but offer longer lifespans and better thermal stability, making them preferred in electric vehicles, energy storage systems, and certain industrial applications where safety and cycle life are paramount.

(e) Cobalt-Free Lithium-Ion Batteries

Cobalt-free lithium-ion batteries, championed by companies such as CATL and Tesla, address concerns of resource consumption and toxicity. The move toward cobalt-free cathode materials is essential to reduce the environmental and ethical concerns associated with cobalt mining. These batteries aim to maintain high energy density while avoiding supply chain issues related to cobalt. This battery technology avoids the supply chain and cost issues related to cobalt and promises improved energy density, cycle life, and safety. The challenge lies in scaling production while maintaining performance.

(f) Metal Hydrogen Batteries

Metal hydrogen batteries, especially solid-state hydrogen storage batteries, boast unparalleled energy capacity, efficiency, and environmental friendliness and address issues of heat dissipation and practical implementation, holding the potential to revolutionize energy storage. However, their development and commercialization are still in progress due to challenges related to hydrogen storage and cycling stability.

(g) Graphene Batteries

Graphene batteries hold promise as a superior alternative to conventional Li-ion batteries due to their unique potential in improving energy storage and charge/discharge rates. Graphene, a single layer of carbon atoms in a hexagonal lattice, is atom-thin yet incredibly robust, possessing superior electrical and thermal conductivity, high surface area, and thermal transparency.

This suite of attributes enables graphene batteries to achieve remarkable energy storage capabilities. The large surface area of graphene accommodates more ions, facilitating a high energy density of up to 1000 Wh/kg, four times greater than that of Li-ion batteries (250 Wh/kg). Consequently, graphene batteries offer substantially more power storage per unit volume or weight.

Owing to its high conductivity, graphene-based batteries charge and discharge exponentially faster than their Li-ion counterparts, reducing charging times from hours to mere seconds. Furthermore, graphene batteries enjoy longer lifespans due to the reduced build-up of lithium deposits on the anode, an issue that causes Li-ion batteries to degrade over time. Current estimates project 3000–5000 cycles for graphene batteries, compared to 500–1000 cycles for Li-ion batteries.

Safety concerns associated with Li-ion batteries, such as fire or explosion risks from flammable liquid electrolytes, are significantly mitigated with graphene batteries due to their use of less flammable, solid-state electrolytes. Moreover, the high thermal conductivity of graphene batteries allows for rapid heat dissipation, reducing overheating and thermal runaway risks.

Environmental advantages are another advantage of graphene batteries, which use nontoxic elements such as lithium, cobalt, or nickel, thereby reducing their ecological impact compared to Li-ion batteries. Additionally, the widespread availability and recyclability of carbon, the building block of graphene, render graphene batteries a more sustainable option, reducing reliance on less abundant elements such as lithium.

Despite the benefits demonstrated in laboratory environments, commercial graphene-based batteries are still under research and development. Current obstacles to commercialization include high production costs, scalability issues, and maintaining the stability and consistency of graphene-based materials.

Producing high-quality graphene at an affordable rate is a major hurdle, given the current cost estimates of approximately $100–300/g using the chemical vapor deposition (CVD) method. Limited infrastructure for large-scale production of graphene batteries, lack of standardization, and concerns about the durability of the thin, fragile graphene material are additional challenges. Furthermore, as a modern technology, graphene batteries may face regulatory obstacles before their widespread application.

(h) Lithium Nickel–Cobalt-Manganese Batteries

Lithium nickel-cobalt-manganese batteries are also known as NMC, Li-NMC, LNMC, or NCM batteries and as one of the two *ternary* (meaning "composed of three parts") batteries. Considered an evolution of traditional lithium-ion batteries, they are already in use in many consumer electronics and electric vehicles, contributing to the ongoing transition from older lithium-ion chemistries. They have a higher energy density and power output than many other types of Li-ion batteries. These batteries combine the high specific energy of nickel, the good stability of manganese, and the easy accessibility of cobalt. They also demonstrate improved safety and life span over older types of Li-ion batteries. However, these batteries are more expensive because of the use of the rare and expensive material cobalt. In addition, there are also ethical concerns regarding cobalt mining and the less environmentally sustainable material manganese. Moreover, they possess a risk of thermal runaway and fire, although this risk is less than that of older types of Li-ion batteries. NCM batteries are produced and consumed globally with a significant presence in Asia, particularly in China, South Korea, and Japan. LG Chem, Panasonic, Samsung SDI, and CATL are some of the leading companies producing these batteries. NCM batteries are primarily used in EVs due to their higher energy density and efficiency. They are also used in energy storage systems (ESS) and some portable electronics.

(i) Lithium-Nickel–Cobalt-Aluminum (NCA) Batteries

Lithium nickel-cobalt-aluminum (also known as NCA) batteries typically have higher energy density and power output than many other types of Li-ion batteries, providing a significant advantage in electric vehicles and other high-energy applications. They combine the high specific energy of nickel, the improved stability of aluminum, and the easy accessibility of cobalt, as well as improved safety and life span over older types of Li-ion batteries. However, the use of cobalt makes these batteries more expensive, as cobalt is a rare and expensive material. There are also ethical concerns regarding cobalt mining. In addition, they also possess a risk of thermal runaway and fire, although this risk is less than that of older types of Li-ion batteries. NCA batteries also have global production and consumption, with a significant presence in the U.S. and Japan. Panasonic and Tesla are the leading companies known to produce NCA batteries. Panasonic produces these batteries primarily for Tesla's electric vehicles. NCA batteries have a high energy density, making them suitable for electric vehicles, including Tesla's line of cars. They are also used in power tools and other high-drain portable devices. Their development is ongoing to further improve their performance and longevity.

(j) Lithium-Polymer Batteries

Lithium-polymer (Li–Po) batteries have been gaining popularity due to their flexible form factor and high energy density. They have a flexible and lightweight form factor, making them suitable for specific applications such as wearables and mobile devices. However, they faced challenges in energy density compared to some other options.

(k) Zinc-Manganese Oxide Batteries

Zinc-manganese oxide (ZMO) batteries are known for their stability, relatively good safety profile, and low cost compared to lithium-ion batteries. They are used in various applications, including consumer electronics and grid storage. However, their lower energy density necessitates overcoming challenges to match the capabilities of lithium-ion batteries.

(l) Lithium-Manganese Oxide Batteries

Lithium manganese oxide (LMO) batteries offer good thermal stability and safety but may have slightly lower energy density compared to other advanced options. They are used in some hybrid vehicles and power tools.

(m) Lithium-Titanate Batteries

Lithium titanate (Li titanate) batteries are known for their high power density and fast charge and discharge rates, along with their long cycle life. They find applications in electric vehicles and specific areas where rapid charging and high power output are critical. However, they are still in the experimental or research phase, and more development is needed.

(n) Organosilicon Electrolyte Batteries

Organosilicon electrolyte batteries bring forth superior safety and stability compared to their lithium-ion counterparts, positioning them for critical applications such as electric vehicles and medical equipment. The absence of flammable liquid electrolytes and reduced hazardous materials make them a beacon of safety.

(o) NanoBolt Lithium Tungsten Batteries

NanoBolt lithium tungsten batteries introduce carbon multilayered nanotubes and tungsten, ushering in improvements in energy storage and rapid charging. This technology holds the promise of longer lasting EVs but needs to overcome challenges in scalability and commercial viability.

(p) Sodium-Ion Batteries

Sodium-ion (saltwater) batteries offer an intriguing sustainable energy storage option. With its low cost, nontoxic nature, and long lifespan, this battery technology presents an economically viable alternative. However, its limitations in energy density and charging cycles require further innovation.

In conclusion, all advanced battery technologies in development symbolize a change in basic assumptions in energy storage, setting the stage for a cleaner, more efficient future. As these innovations address challenges in energy density, safety, and sustainability, they also encounter hurdles related to production costs, scalability, and technological maturity. The journey toward the realization of these technologies involves a delicate dance between innovation and practicality, with each advancement propelling us closer to a greener, electrified world.

6. Long-Duration Energy Storage

Long-duration energy storage is becoming increasingly important as the world shifts toward greater reliance on renewable energy sources such as wind and solar power. Unlike traditional fossil fuel-based power plants, power generation and supply from intermittent renewable energy sources can fluctuate based on weather conditions, which causes difficulties in maintaining a stable and reliable energy grid. With long-duration energy storage, excess energy generated during periods of high energy generation can be stored and used during periods of low energy generation, helping to balance energy supply and demand and ensure a stable and reliable energy supply system. It can provide backup power during disaster-related and other power outages or emergencies, ensuring that critical infrastructure and essential services can continue to function. It helps integrate renewable energy sources into the grid by providing a more stable and reliable source of energy, reducing the need for fossil fuel-based backup power plants. It also contributes to time-shift energy use, allowing energy to be stored during off-peak hours when power is cheaper and used during peak hours when power is more expensive.

In on-grid applications, long-duration energy storage can play a significant role in supporting the reliability and resilience of the grid. Energy storage can help to manage peak demand, reduce the need for expensive grid infrastructure upgrades, and provide backup power in case of outages. Long-duration energy storage can be particularly useful in supporting the integration of renewable energy sources, such as wind and solar power, which may generate excess energy at certain times and require storage for later use.

In off-grid applications, such as remote or underserved communities or islands, long-duration energy storage can be a critical component of a sustainable and reliable energy system. These systems can help balance supply and demand, ensure that energy is available when needed, and help expand the generation and consumption of renewable energy in those areas, reducing their reliance on costly and environmentally damaging diesel generators.

The levelized cost of storage for long-duration stationary applications is expected to be reduced to $0.05/kWh in 2030, accounting for a reduction of 90% from 2020 baseline costs. Meeting this levelized cost target, energy storage can achieve commercial viability for a wide range of applications, such as meeting peak demand, preparing grids for fast charging of electric vehicles, and ensuring the reliability of critical infrastructures, including communications and information technology. At the same time, the manufactured cost for a battery pack is expected to fall to $80/kWh by 2030 for a 300-mile range electric vehicle, which represents a 44% reduction from the current cost of $143 per rated kWh. Meeting this cost target would help dramatically expand a cost-competitive electric vehicle market.

7. Hydrogen Solutions

In the relentless pursuit of a fully renewable energy future, hydrogen fuel cells emerge as a beacon of hope, offering innovative solutions to the challenges that lie ahead. Through remarkable technological breakthroughs encompassing materials, designs, and manufacturing techniques, hydrogen fuel cells are on the cusp of redefining energy storage applications in vehicle transportation and power generation. The diverse array of fuel cell types, including proton exchange membrane fuel cells (PEMFCs), solid oxide fuel cells (SOFCs), and molten carbonate fuel cells (MCFCs), each tailored for specific applications and operating conditions, paints a comprehensive picture of the possibilities that lie before us.

Proton exchange membrane (PEM) fuel cells, born in the 1960s, have matured to find applications across industries. Leveraging platinum-based catalysts and polymer electrolyte membranes, they exemplify the cornerstone of modern fuel cell technology, efficiently converting hydrogen into power.

The horizon is broadening, with advanced fuel cells leading the charge. These innovations, armed with novel materials and designs, hold the promise of pushing performance and efficiency boundaries. Novel catalysts alleviate reliance on costly elements such as platinum, while enhanced membrane materials enhance durability in challenging environments.

As the hydrogen fuel cell industry navigates the path toward a sustainable future, overcoming formidable challenges is paramount. Extending fuel cell lifetimes from the present 3–5 years, coupled with potential performance degradation, to a robust 5–10 years with minimal loss is a pivotal stride forward. Likewise, elevating fuel cell efficiency from the current 40–60% range to a competitive 60% is essential, aligning it with energy technologies such as natural gas turbines.

In terms of economics, the quest for market viability takes center stage. Slashing the cost of hydrogen storage and transportation from $3–10/kg to below $2/kg becomes the bridge that links us to conventional fossil fuels. An investment exceeding $100 billion in infrastructure, encompassing the establishment of hydrogen production, storage, and distribution facilities, underscores our commitment to a sustainable future.

Safety also commands the attention of the hydrogen industry. While hydrogen boasts relative safety, addressing concerns linked to its production, transportation, and utilization is vital to instilling public confidence in fuel cell technology.

The rise of green hydrogen will be a transformative force in energy expansion, climate change mitigation, and energy security. This versatile energy storage medium holds the key to integrating renewable sources into the grid and driving the decarbonization of challenging industries such as steel and cement production.

Moreover, hydrogen's application in fuel cells across transportation, heating, and industrial processes presents an impactful avenue for reducing greenhouse gas emissions, propelling us toward net-zero targets. Additionally, by diversifying energy sources through hydrogen production from renewables such as wind and solar energy, we bolster energy security and lessen our dependence on fossil fuels.

In addition, a global initiative takes shape, marked by a multitude of hydrogen projects in various stages of development. According to the International Energy Agency's Clean Energy Database, there are 1477 hydrogen projects worldwide in various stages. A wide range of technologies and applications are developed in these projects. These endeavors signify the collective drive toward a cleaner, more resilient energy future.

Figure 7.17 highlights the potential for hydrogen to play a significant role in the global transition to a low-carbon economy. It shows that the majority of the hydrogen projects are in the feasibility study (521) and concept (333) stages. These projects are followed by projects in operation (227), DEMO (191), and final investment (123) decision stages. Regarding technology, 333 projects use alkaline electrolysis technology, which is the most widely used technology. This group is followed by 291 PEM and 793 other electrolysis projects. In addition, there are 125 fossil fuel-based hydrogen projects, primarily natural gas, with carbon capture, utilization, and storage (CCUS) projects.

The focuses of these hydrogen projects vary depending on the technology used. The projects using alkaline electrolysis and other electrolysis are likely focused on producing green hydrogen using renewable energy sources such as wind and solar power. PEM projects, on the other hand, may be focused on producing hydrogen for fuel cells in transportation or stationary applications. Natural gas with CCUS projects may be focused on producing blue hydrogen, which is produced

Technology	Concept	DEMO	Feasibility study	Final investment decision	Under construction	Operational	Decommisioned	Other/Unknown	Subtotal
ALK	8	46	32	21	11	73	2	2	195
PEM	15	73	49	31	29	94	0	0	291
SOEC	6	18	5	3	2	4	0	0	38
Other Electrolysis	272	44	353	61	26	34	0	3	793
Biomass w CCUS	1	0	1	0	0	0	0	0	2
Biomass	3	7	6	3	2	4	0	0	25
Coal w CCUS	1	0	3	0	0	4	0	0	8
NG w CCUS	24	1	62	3	1	9	0	1	101
Oil w CCUS	0	0	3	0	0	5	0	0	8
Other	3	2	7	1	1	0	1	1	16
Subtotal	333	191	521	123	72	227	3	7	1477

Fig. 7.17 Hydrogen projects in the world. *Data Source* IEA Clean Energy Database

from natural gas with carbon capture and storage technologies. Finally, there are a small number of biomass, coal, and oil CCUS projects that may focus on producing hydrogen from these sources while capturing and storing the resulting carbon emissions. These hydrogen projects represent a significant global effort to develop and deploy hydrogen technologies across energy storage for a range of applications from grid support to transportation.

Summary

In the context of achieving net-zero carbon emissions, energy storage is becoming increasingly crucial to ensure that renewable energy sources such as solar and wind power can be utilized to their fullest potential. Two of the most prevalent technologies for utility-scale energy storage are PHS and Li-ion batteries. While both technologies have their respective advantages and disadvantages, finding the most viable solutions for energy storage in the future will be critical to the success of the net-zero carbon emission initiative.

PHS utilizes gravity to store energy by pumping water from a lower reservoir to a higher reservoir during periods of low demand for electricity. This water is then released to flow back down through a turbine when demand is high, generating electricity. Although PHS is a mature technology that has been in use for several decades, there are still significant challenges, such as significant upfront capital costs, large and unique land-use requirements, and legal and regulatory requirements, making this technology less attractive for accelerated deployment. However, research results reveal that PHS has a much longer life cycle of 80 or more years than the 10-year life cycle of Li-ion batteries and that closed-loop PHS is an ideal and sustainable technology to meet the large-scale and long-term energy storage demand. Additionally, the use of AI technologies and sustainability metrics and tools for PHS resource mapping and sustainable PHS planning also provides viable solutions for the further expansion of PHS technology.

On the other hand, Li-ion batteries are a newer technology that has seen rapid growth in recent years, largely due to the boom in electric vehicles. They are ideal for short-term and small-scale energy storage and are increasingly scalable. Additionally, they have a smaller physical footprint and are more flexible in terms of their placement. However, the existing batteries have a much shorter lifespan and require frequent replacement, leading to high operating costs. They are also susceptible to thermal runaway and other safety hazards.

Given the advantages and disadvantages of both technologies, finding the most viable solutions for energy storage in the future will depend on a variety of factors, including the specific energy storage needs of the utility company, the geographic location, and the availability of resources.

One potential solution to address the high capital cost of PHS is technological innovation. For example, the State Grid Corporation in China has developed a new type of PHS technology that uses seawater as the storage medium. This technology is expected to significantly reduce capital costs compared to traditional PHS.

Another potential solution is the development of economies of scale, which can also help reduce the upfront cost of PHS.

Similarly, technological advancements in Li-ion battery technology can help improve its lifespan and safety while reducing its operating costs. Research is currently underway to develop new battery chemistries that can provide longer-lasting, safer, and more sustainable energy storage. Additionally, recycling and repurposing used batteries can help reduce the environmental impact of Li-ion battery production and disposal.

In terms of the net-zero carbon emission initiative, both PHS and Li-ion batteries have a role to play. PHS can provide long-term energy storage for larger-scale renewable energy projects, while Li-ion batteries can provide short-term energy storage and be utilized in smaller-scale renewable energy projects. Finding the right balance between these two technologies will be crucial in achieving the net-zero carbon emission goal.

Overall, the challenges of expanding energy storage capacities underscore the need for continued research, development, and investment in long-duration energy storage technologies to overcome these challenges and fully realize their potential as a critical component of a sustainable and reliable energy grid.

In conclusion, PHS, Li-ion batteries and other advanced energy storage technologies have common and unique advantages and disadvantages, but finding viable solutions for energy storage in the future will depend on a variety of factors. While technological innovation and economies of scale can help reduce the high capital cost of PHS, advancements in Li-ion battery technology and other advanced energy storage technologies can improve their efficiency, lifespan, safety, and operating costs. Ultimately, finding the right balance between these two technologies will be crucial in achieving the net-zero carbon emission goal.

Activities

Further Readings

1. U.S. EPA (2022). Electricity Storage. https://www.epa.gov/energy/electricity-storage
2. International Hydropower Association (2022). Pumped hydro. https://www.hydropower.org/factsheets/pumped-storage
3. U.S. Department of Energy (2022). Pumped Storage Hydropower. https://www.energy.gov/eere/water/pumped-storage-hydropower
4. Rojanasakul, M. & Bearak, M. (2023). Is It a Lake, or a Battery? A New Kind of Hydropower Is Spreading Fast (video). https://www.nytimes.com/interactive/2023/05/02/climate/hydroelectric-power-energy.html
5. Documentary of the construction of the Limmern PSP (video). https://www.youtube.com/watch?v=Hv8DPzaed0Q

6. Michigan Technological University (2022). Using Old Mines for Pumped Hydropower Energy Storage is a Game-Changer (video). https://www.youtube.com/watch?v=haqC3beaRUM
7. UN Climate Technology Centre & Network (2011). Compressed Air Energy Storage (CAES). https://www.ctc-n.org/technologies/compressed-air-energy-storage-caes
8. Geology, Energy & Minerals Science Center (2023). Geologic Energy Storage. https://www.usgs.gov/media/images/geologic-energy-storage
9. UN Climate Technology Centre & Network (2007). Flywheels. https://www.ctc-n.org/technologies/flywheels
10. Ugoji, E. (2023). 10 Battery Technologies Currently in Development. https://www.topspeed.com/battery-technologies-currently-in-development/#silicon-anode-lithium-ion-batteries
11. BloombergNEF (2023). 10 Energy Storage Trends in 2023. https://about.bnef.com/blog/top-10-energy-storage-trends-in-2023/
12. Crownhart, C. (2023). What's next for batteries? https://www.technologyreview.com/2023/01/04/1066141/whats-next-for-batteries/
13. Sarkis, S. (2023). Ternary lithium battery. https://www.roypowtech.com/blog/are-lithium-phosphate-batteries-better-than-ternary-lithium-batteries
14. Reid, M. (2023). Sodium-ion batteries: disrupt and conquer? https://www.woodmac.com/news/opinion/sodium-ion-batteries-disrupt/
15. Crownhart, C. (2023). How sodium could change the game for batteries. https://www.technologyreview.com/2023/05/11/1072865/how-sodium-could-change-the-game-for-batteries/
16. CNBC (2023). How Sodium-Ion Batteries May Challenge Lithium (video). https://www.youtube.com/watch?v=RQE56ksVBB4
17. Tech Space (2022). Graphene Batteries FINALLY Hit the Market! https://www.youtube.com/watch?v=zTrvPqU0rkE
18. Maia, B. (2023). What Nobody Is Telling You About Solid-State Batteries. https://www.topspeed.com/nobody-telling-about-solid-state-batteries/
19. Sleppy, J. (2023). Batteries vs. Supercapacitors? The Answer is Both. https://www.capacitechenergy.com/blog/batteries-vs-supercapacitors-the-answer-is-both
20. Clifford, C. (2022). Hydrogen power is gaining momentum, but critics say it's neither efficient nor green enough. https://www.cnbc.com/2022/01/06/what-is-green-hydrogen-vs-blue-hydrogen-and-why-it-matters.html
21. National Grid (2023). The hydrogen colour spectrum. https://www.nationalgrid.com/stories/energy-explained/hydrogen-colour-spectrum
22. Celsius Wiki (2018). Thermal Energy Storage. https://celsiuscity.eu/thermal-energy-storage/

Closed Questions

1. What are the main types of energy storage technologies currently available?
   (a) Mechanical, electrical, electrochemical, and thermal technologies are available
   (b) Energy storage technologies are limited to batteries only
   (c) Pumped hydro storage is the only available type of energy storage technology
   (d) None of these technologies exist.
2. What is the concept of pumped hydro storage and its significance in energy storage?
   (a) Pumped hydro storage has no significance in energy storage
   (b) Pumped hydro storage is a concept related to water conservation
   (c) Pumped hydro storage is a method to store energy using gravitational potential energy
   (d) Pumped hydro storage involves using compressed air to store energy.
3. What role do compressed air energy storage (CAES) systems play in storing energy?
   (a) CAES systems do not contribute to energy storage
   (b) CAES systems use compressed air to store energy in underground reservoirs
   (c) CAES systems are used to store energy in the form of batteries
   (d) CAES systems store energy through photosynthesis.
4. What are the potential economic advantages of integrating energy storage into renewable energy systems?
   (a) There are no economic advantages to integrating energy storage with renewable energy
   (b) Integrating energy storage with renewable energy increases electricity costs
   (c) Energy storage has no impact on renewable energy systems
   (d) Energy storage integration helps reduce off-peak energy waste, peak demand, and electricity costs.
5. What role do energy storage technologies play in enhancing the integration of intermittent renewable energy sources?
   (a) Energy storage technologies have no impact on the integration of renewable energy sources
   (b) Energy storage technologies cause instability in renewable energy integration
   (c) Energy storage technologies help store excess renewable energy for later use
   (d) Energy storage technologies lead to the complete shutdown of renewable energy sources.
6. What energy storage technologies have limitations in terms of scalability and duration of storage?
   (a) Batteries and hydrogen face limitations in scalability and storage duration

(b) All energy storage technologies are highly scalable and have unlimited storage duration
(c) Batteries have no limitations in scalability and storage duration
(d) Energy storage technologies are limited to small-scale applications only.

7. What are the safety concerns related to energy storage systems, especially in residential and commercial settings?
   (a) Energy storage systems pose no safety risks in residential and commercial settings
   (b) Safety concerns include potential fires, toxic emissions, and explosions
   (c) Energy storage systems are completely safe and risk-free
   (d) Safety concerns only apply to large-scale energy storage systems.
8. What are the societal benefits of the widespread adoption of energy storage technologies?
   (a) There are no societal benefits associated with energy storage technologies
   (b) Energy storage technologies lead to increased pollution and harm to society
   (c) Energy storage technologies have no impact on society
   (d) The widespread adoption of energy storage technologies enhances grid stability and reliability.
9. What is the potential for energy storage to enable electrification of transportation and its associated benefits?
   (a) Energy storage has no potential to enable electrification of transportation
   (b) The electrification of transportation has negative impacts on the environment
   (c) Energy storage plays a crucial role in enabling decarbonized electrification of transportation and reducing emissions and dependence on fossil fuels
   (d) The electrification of transportation relies solely on traditional fuels.
10. What are the key challenges hindering the widespread deployment of energy storage solutions?
   (a) The key challenges include high upfront costs, policy barriers, and technical limitations
   (b) There are no challenges in deploying energy storage solutions
   (c) Energy storage solutions are readily available and have no deployment challenges
   (d) Energy storage solutions are easy to deploy with no associated challenges.
11. What are the opportunities for decentralized energy storage solutions, and how do they benefit the overall energy landscape?
   (a) Decentralized energy storage solutions have no benefits for the overall energy landscape
   (b) Decentralized energy storage solutions contribute to increased grid dependency
   (c) There are no opportunities for decentralized energy storage solutions

   (d) Decentralized energy storage solutions provide grid flexibility and enhanced energy security.
12. What is the importance of standardization and policy support in accelerating the adoption of energy storage technologies?
   (a) Standardization and policy support have no impact on the adoption of energy storage technologies
   (b) Energy storage technologies are already widely adopted without any standardization or policy support
   (c) Standardization and policy support create a favorable environment for investment and innovation in energy storage
   (d) Standardization and policy support hinder the growth of energy storage technologies.
13. What are the safety concerns related to energy storage systems, especially in residential and commercial settings?
   (a) Energy storage systems pose no safety risks in residential and commercial settings
   (b) Safety concerns include potential fires, toxic emissions, and explosions
   (c) Energy storage systems are completely safe and risk-free
   (d) Safety concerns only apply to large-scale energy storage systems.
14. What are the economic challenges of energy storage, such as high upfront costs and uncertain revenue streams?
   (a) No economic challenges are associated with energy storage
   (b) Energy storage technologies are low-cost and provide stable revenue streams
   (c) Economic challenges include high upfront costs and uncertainty in revenue generation
   (d) Energy storage has no impact on the economy.
15. What are the ongoing research and development efforts to improve energy storage technologies?
   (a) Ongoing research aims to improve the efficiency and capacity and reduce the costs of energy storage technologies
   (b) There are no ongoing research and development efforts for energy storage technologies
   (c) Energy storage technologies are already perfected and do not require further research
   (d) Research efforts focus solely on decreasing the use of energy storage technologies.
16. How can advancements in materials science lead to better-performing and more efficient batteries?
   (a) Advancements in materials science have no impact on battery performance
   (b) Advancements in materials science can actually hinder battery efficiency
   (c) Advancements in materials science may lead to better-performing and more efficient batteries
   (d) Advancements in materials science only affect nonenergy storage technologies.

17. What is the potential of using machine learning and artificial intelligence to optimize energy storage operations?
    (a) Machine learning and artificial intelligence cannot be applied to energy storage operations
    (b) Machine learning and artificial intelligence have the potential to optimize energy storage operations
    (c) Machine learning and artificial intelligence may lead to worse performance in energy storage operations
    (d) Machine learning and artificial intelligence are only used in nonenergy-related fields.
18. What are some innovative approaches to overcome the intermittency issues of renewable energy through storage solutions?
    (a) There are no innovative approaches to address the intermittency issues of renewable energy
    (b) Innovative approaches cannot effectively address the intermittency issues of renewable energy
    (c) Innovative approaches only apply to nonrenewable energy sources
    (d) Innovative approaches exist to overcome the intermittency issues of renewable energy through storage solutions.
19. How can the integration of various energy storage technologies NOT create hybrid systems with improved performance?
    (a) Integrating energy storage technologies does not affect the performance of hybrid systems
    (b) Integrating energy storage technologies may lead to hybrid systems with worse performance
    (c) The integration of various energy storage technologies can create hybrid systems with improved performance
    (d) Hybrid systems with improved performance can only be achieved through the nonintegration of energy storage technologies.
20. What is the concept of "second-life" battery applications NOT and their role in extending the usefulness of batteries?
    (a) Second-life battery applications do not exist and have no role in extending battery usefulness
    (b) Second-life battery applications have no role in extending the usefulness of batteries
    (c) The concept of "second-life" battery applications is only related to non-battery technologies
    (d) Second-life battery applications involve reusing car batteries after their automotive life with a residual capacity of approximately 70–80% for other effective purposes to extend their usefulness.

Open Questions

1. How does battery energy storage work, and what types of batteries are used?
2. How do thermal energy storage systems function, and what are their applications?
3. What are the principles behind flywheel energy storage technology and its advantages.
4. What are supercapacitors, and how do they differ from traditional batteries in energy storage?
5. How does hydrogen energy storage work, and what potential does it hold for large-scale applications?
6. How do energy storage technologies contribute to grid stability and reliability?
7. How can energy storage technologies help reduce peak demand and electricity costs?
8. How can energy storage provide backup power and support critical infrastructure during emergencies?
9. What are the current trends in the global energy storage market?
10. Which regions or countries have made significant progress in deploying energy storage technologies, and what factors contributed to their success?
11. How has the adoption of energy storage technologies influenced the growth of renewable energy installations?
12. Which sectors (e.g., residential, commercial, industrial) are driving the demand for energy storage technologies the most?
13. How do government policies and regulations impact the market development of energy storage?
14. What are some successful case studies of energy storage projects that have been implemented at scale?
15. How do different financing models influence the deployment of energy storage technologies?
16. What are the main technical challenges faced by energy storage technologies?
17. How do issues such as self-discharge, capacity fading, and thermal management affect battery energy storage systems?
18. What are the environmental impacts associated with the production and disposal of energy storage technologies?
19. What challenges arise from integrating energy storage systems into existing power grids?
20. How do extreme weather conditions or temperature variations affect the performance of energy storage technologies?

Appendix

P. Yang, *Renewable Energy*, https://doi.org/10.1007/978-3-031-49125-2

Best Research-Solar-Cell Efficiencies

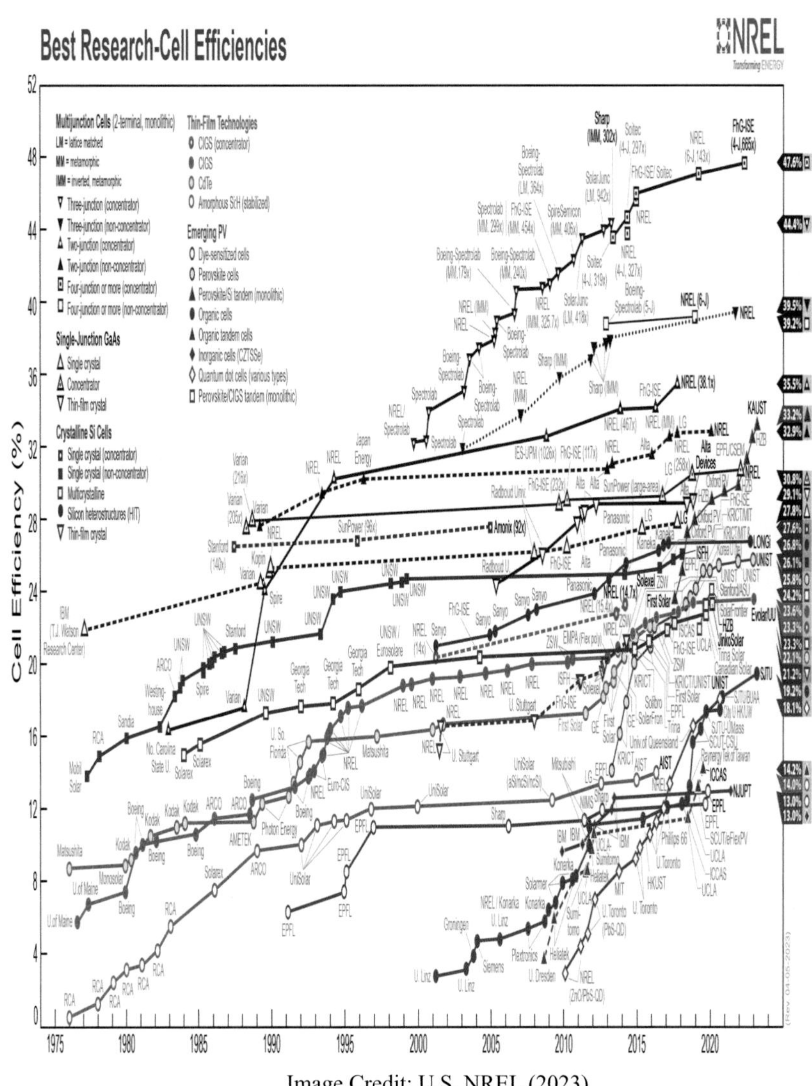

Image Credit: U.S. NREL (2023)

Solar PV: Existing and Necessary Future Installed Solar PV Capacities

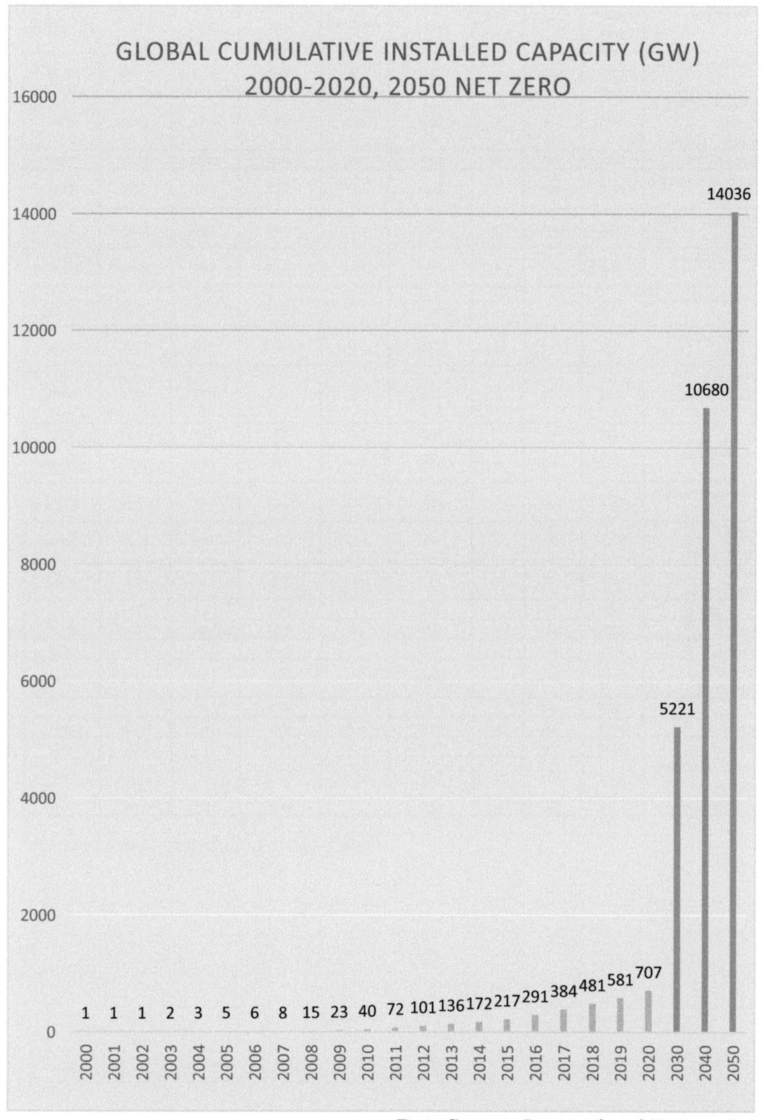

Data Source: International Energy Agency

Costs of Future Power Generation Technologies

Technology	First available year	Size (MW)	Lead time (years)	Base overnight cost (2021$/kW)	Techno-logical optimism factor	Total overnight cost (2021$/kW)	Variable O&M (2021 $/MWh)	Fixed O&M (2021$/ kW-y)	Heat rate (Btu/kWh)
Ultra-supercritical coal (USC)	2025	650	4	4074	1	4074	4.71	42.49	8638
USC with 30% carbon capture and sequestration (CCS)	2025	650	4	5045	1.01	5096	7.41	56.84	9751
USC with 90% CCS	2025	650	4	6495	1.02	6625	11.49	62.34	12507
Combined-cycle—single-shaft	2024	418	3	1201	1	1201	2.67	14.76	6431
Combined-cycle—multi-shaft	2024	1083	3	1062	1	1062	1.96	12.77	6370
Combined-cycle with 90% CCS	2024	377	3	2736	1.04	2845	6.11	28.89	7124
Internal combustion engine	2023	21	2	2018	1	2018	5.96	36.81	8295
Combustion turbine—aeroderivative	2023	105	2	1294	1	1294	4.92	17.06	9124
Combustion turbine—industrial frame	2023	237	2	785	1	785	4.71	7.33	9905
Fuel cells	2024	10	3	6639	1.09	7224	0.62	32.23	6469
Nuclear—light water reactor	2027	2156	6	6695	1.05	7030	2.48	127.35	10443
Nuclear—small modular reactor	2028	600	6	6861	1.1	7547	3.14	99.46	10443
Distributed generation—base	2024	2	3	1731	1	1731	9.01	20.27	8923
Distributed generation—peak	2023	1	2	2079	1	2079	9.01	20.27	9907
Battery storage	2022	50	1	1316	1	1316	0	25.96	NA
Biomass	2025	50	4	4524	1	4525	5.06	131.62	13500
Geothermal	2025	50	4	3076	1	3076	1.21	143.22	8813
Conventional hydropower	2025	100	4	3083	1	3083	1.46	43.78	NA
Wind	2024	200	3	1718	1	1718	0	27.57	NA
Wind offshore	2025	400	4	4833	1.25	6041	0	115.16	NA
Solar thermal	2024	115	3	7895	1	7895	0	89.39	NA
Solar photovoltaic (PV) with tracking	2023	150	2	1327	1	1327	0	15.97	NA
Solar PV with storage	2023	150	2	1748	1	1748	0	33.67	NA

Data Source: U.S. Department of Energy

Terms of Renewable Energy

Aquaculture: The farming of organisms that live in water, such as fish, shellfish, and algae, for commercial purposes.

Baseload capacity: The maximum power output that power generating equipment can deliver consistently over a specified period of time to meet the base load demand of a system.

Baseload plant: A power plant, typically equipped with high-efficiency steam-electric units, which is operated to provide a sizable portion or the entirety of the minimum load of a system. Baseload plants run continuously and produce power at a constant rate to maximize efficiency and minimize operating costs.

Baseload: The minimum level of electric power demand required to meet the continuous, steady demand over a specific time period.

Binary-cycle plant: A geothermal power plant that utilizes a closed-loop heat exchange system. The heat from geothermal fluid is transferred to a lower-boiling-point fluid, which vaporizes and drives a turbine/generator set.

Biomass: Energy sources derived from organic matter such as wood, agricultural residues, and living-cell materials. Biomass can be converted into heat energy through processes such as combustion, digestion, pyrolysis, and gasification.

Capability: The maximum load that a power generating unit, power plant, or electrical apparatus can carry under specific conditions for a given time period without exceeding approved limits of temperature and stress.

Capacity factor: A measure, usually expressed as a percentage, which represents the ratio of actual energy production of a power plant or energy-generating system to its maximum possible output capacity.

Capacity: The maximum amount of energy that an energy plant or energy-generating system can produce over a specified period, typically measured in megawatts (MW) or gigawatts (GW), as rated by the manufacturer.

Carbon dioxide: A colorless, odorless gas that is a normal component of the Earth's atmosphere. Carbon dioxide is exhaled by humans and animals and absorbed by plants and the ocean.

Combined cycle: A power generation technology that utilizes waste heat from gas turbines by routing it to a boiler or heat recovery steam generator to produce steam for a steam turbine, increasing overall efficiency.

Condensate: Water formed by condensation of steam.

Cooling tower: A structure used in water-cooled heat extraction systems to facilitate the efficient cooling of hot water. The tower operates by spraying the hot water from the top and allowing it to cascade down in the presence of an upward airflow, which causes a portion of the water to evaporate and lower the overall temperature.

Cost: The amount of money paid to acquire resources, such as plant and equipment, fuel, or labor services.

Crust: Earth's outermost layer of solid rock, also known as the lithosphere.

Direct use: The utilization of solar or geothermal heat directly for purposes such as space heating and cooling, food preparation, and industrial processes, without converting it to electricity.

Dispatch: The operational control of an integrated power system to allocate generation to specific plants and sources of supply to provide reliable and cost-effective power as the load fluctuates. It also involves the control of high voltage lines, substations, equipment, safety procedures, and energy transactions with other utilities.

Distribution: The process of delivering power to end-use consumers, such as residential, commercial, and industrial customers.

Drilling: The process of boring into the Earth to access geothermal resources, typically using specialized drilling equipment adapted for geothermal applications.

Dry steam: Superheated steam at a hot temperature that does not contain liquid water.

Economies of scale: The phenomenon where the average cost of producing each unit of output decreases as the scale of production increases. This reduction in cost is achieved through operational efficiencies, such as spreading fixed costs over a larger production volume, taking advantage of volume discounts from suppliers, and benefiting from specialized equipment and technology.

Electric utility: An authorized entity, such as a corporation, person, agency, authority, or instrumentality, entrusted with the generation, transmission, distribution, or sale of electricity predominantly for public consumption. Electric utilities own and manage infrastructures, including power plants, transmission lines, and distribution systems, for the purpose of supplying electricity to customers. However, cogenerators and small power producers, who produce electricity primarily for their own usage and occasionally export surplus power to the grid, are not classified as electric utilities.

Emissions standard: The legally permitted maximum amount of a pollutant that can be released from a particular source into the environment.

Energy demand: The rate at which energy is required or consumed by a system, part of a system, or a piece of equipment, either at a specific moment or averaged over a designated time period.

Energy source: A primary source of energy that is converted into power, heat, solid or liquid fuel, or other forms of energy through electromagnetic, chemical, mechanical, or other means. Common energy sources include sunlight, wind, water, biomass, geothermal, natural gas, oil, coal, and uranium, among others.

Energy: The capacity to perform work, which can exist as potential energy (stored energy) or kinetic energy (energy of motion). Energy exists in various forms and can be converted from one form to another to perform useful work. Currently, a massive portion of the world's energy is derived from burning fossil fuels, which produce heat that is utilized through mechanical or other processes to accomplish tasks.

Environmental impact study: A comprehensive document required by international standards and national laws, accompanying project and program proposals that have the potential to cause substantial environmental impacts and affect the surrounding areas. The study assesses and analyzes the potential effects on the environment, providing information to aid in decision-making and mitigation strategies.

Facility: A site where energy generation and conversion take place, housing generators and equipment that facilitate the conversion of energy from one form to another. These facilities encompass various energy generation methods, including thermal, mechanical, chemical, nuclear, solar PV, and bioenergy processes. Within a facility, multiple generators may be present, employing different methods to convert energy. In the case of cogeneration, the facility integrates an industrial or commercial process with energy generation. Overall, a facility acts as a central hub where different energy technologies converge to produce a variety of usable energy outputs.

Flash steam: Steam generated when the pressure on a geothermal liquid is rapidly reduced, resulting in a phase change from liquid to steam.

Fossil fuel: Any naturally occurring fuel derived from organic materials, such as coal, petroleum, and natural gas.

Fossil-fuel plant: A plant that utilizes coal, petroleum, or gas as its primary source of energy.

Fuel: Any substance that can undergo combustion to release heat energy; nuclear materials that can undergo fission in a chain reaction to produce heat.

Generating unit: A combination of physically interconnected generator(s), reactor(s), boiler(s), combustion turbine(s), or other prime mover(s) that operate collectively to generate electric power.

Geology: The scientific study of the Earth's composition, structure, natural processes, and history.

Geophysical survey: Geophysical techniques employed during exploration and drilling stages to identify resources and determine optimal locations for drilling production wells. These methods may encompass gravity surveys, ground magnetic surveys, magnetotelluric surveys, electrical resistivity surveys, and seismic surveys.

Geothermal energy: The utilization of the Earth's internal heat for power generation or direct applications such as space heating or industrial steam.

Geothermal heat pumps: Heat pump systems that harness the consistent temperature of the Earth's interior as a renewable energy source for both heating and cooling purposes. In cooling mode, heat is withdrawn from the space and discharged into the Earth, while in heating mode, heat is extracted from the Earth and supplied to the space.

Geothermal plant: A power plant where the primary mechanism for generating electricity is a turbine driven either by natural steam, which derives its energy from heat in rocks, or by steam produced from hot water or pressurized reservoirs at various depths beneath the Earth's surface.

Geothermal steam: Steam extracted from deep within the Earth's crust is typically used as a source of energy in geothermal power plants.

Geothermal: Relating to the heat generated within the Earth's interior.

Geyser: A natural spring characterized by periodic eruptions of hot water and steam into the air.

Gigawatt (GW): A unit of power equal to one billion (10^9) watts.

Gigawatt hour (GWh): A unit of energy equal to one billion watt-hours.

Greenfield project: A project that involves the development of new infrastructure on a previously unused or undeveloped site, specifically referring to power plants in this context.

Greenhouse effect: The greenhouse effect is the process by which certain atmospheric gases (including carbon dioxide, methane, nitrous oxide, ozone, and chlorofluorocarbon) trap and absorb infrared radiation, resulting in an increase in the Earth's surface temperature.

Grid: A project that involves the development of new infrastructure on a previously unused or undeveloped site, specifically referring to power plants in this context.

Gross generation: The total amount of electric energy produced by the generating units at a power plant or collection of power plants, measured at the terminals of the generators.

Heat exchanger: A device designed to transfer thermal energy from one fluid to another fluid, typically without the fluids coming into direct contact.

Heat pumps: Devices, including geothermal heat pumps, which utilize the temperature difference between a heat source and a heat sink to transfer heat energy for heating or cooling purposes.

Heat Organic Rankine Cycle: A thermodynamic cycle, also referred to as HeatORC, used to convert heat energy into mechanical work using an organic fluid as the working medium; a variation of the more well-known Rankine cycle, which is commonly used in steam power plants.

Hot dry rock: A type of geothermal resource formed when impermeable subsurface rock formations, often granite rock located 5000 km or more below the Earth's surface, are heated by geothermal energy. Hot dry rock resources are being explored as a potential source for energy production.

Hydropower plant: A power plant where the turbine generators are driven by the kinetic energy of falling or flowing water.

Independent power producers (IPPs): Producers characterized by their ability to generate power independently and sell it to the grid or other buyers. Unlike traditional utilities, IPPs typically do not own or operate transmission infrastructure.

Indirect impacts: Economic consequences that affect industries indirectly associated with power plant construction, operation, and maintenance. These impacts quantify the ripple effects on industries providing goods and services to the primary power plant sector due to changes in its activities.

Induced impacts: Impacts that occur when industries, influenced by both direct and indirect effects, modify their employment levels, leading to changes in regional income. These income effects drive further economic consequences, as increased demand for goods and services triggers additional cycles of indirect and induced impacts, promoting growth in other industries.

Injection well: Wells used to inject brine or spent geothermal fluids back into the reservoir after utilizing them in the power production process.

Injection: The process of returning spent geothermal fluids or brine to the subsurface, also known as reinjection.

Kilowatt (kW): A unit of power equal to one thousand (10^3) watts.

Kilowatt-hour (kWh): A unit of energy equal to one thousand watt-hours.

Kinetic energy: The energy possessed by an object due to its motion.

Lead time: The duration between placing an order and receiving the goods or services ordered.

Lease: A contractual agreement between a lessor and a lessee, allowing the lessee to use a vehicle or other property for a specified period and payment, subject to agreed-upon terms and limitations.

Levelized cost of energy (LCOE): A financial metric used to compare the lifetime costs of different energy-generating technologies. It considers all costs associated with constructing and operating an energy-generating system and converts them into an equal annual payment. LCOE enables a fair comparison of energy technologies, regardless of their initial cost or expected lifetime. It is typically expressed in real currency, adjusted to remove the impact of inflation, and is useful for evaluating the economic feasibility of different energy options.

Lithologies: The characteristics and properties of a rock formation, such as its composition, texture, and structure.

Load (electric): The amount of electric power delivered to or required at specific points in an electrical system. The demand originates from the energy-consuming equipment of consumers.

Magma: Molten rock and other materials existing beneath the Earth's crust, generated by the heat of the Earth's interior. Magma can reach temperatures of up to 1200 °C.

Magnetotellurics (MT): A passive geophysical method which uses natural time variations of the Earth's magnetic and electric fields to measure the electrical resistivity of the sub-surface.

Mantle: The Earth's layer of molten rock located beneath the crust and above the core of liquid iron and nickel.

Mechanical efficiency: The ratio of useful output work or power to the input work or power in a mechanical system, a measure of how effectively a machine or device converts input energy into useful output energy, considering losses and inefficiencies within the system.

Megawatt (MW): A unit of power equal to one million (10^6) watts.

Megawatt hour (MWh): A unit of energy equal to one million watt-hours.

Mitigation: Measures, both structural and nonstructural, undertaken to reduce or limit the adverse impacts of natural hazards, environmental degradation, and technological hazards.

Municipal utility: A publicly owned utility that provides water, electricity, or other services to customers within a specific geographic area, such as a city or town. These utilities are owned and operated by the local government or a municipal corporation and are responsible for providing reliable and affordable services to their customers. The regulation of municipal utilities varies across countries but is typically under the authority of local government entities rather than national- or subnational-level agencies.

Natural gas combined cycle: A power-generating technology that combines two cycles—the gas cycle and the steam cycle—to achieve higher efficiency. In the gas cycle, gas is burned in a gas turbine to generate power, while the waste heat from the gas turbine is utilized in the steam cycle to produce steam, which drives a steam turbine to generate additional power. The integration of these cycles maximizes the energy utilization from the fuel, resulting in improved efficiency and lower emissions compared to traditional power plants.

Natural gas: A mixture of hydrocarbon gases, primarily methane, along with nonhydrocarbon gases, found in porous geological formations beneath the Earth's surface. The composition of natural gas can vary depending on its source and location. It can be extracted from both conventional and unconventional reservoirs and is often found in association with crude oil.

Net generation: The amount of electric power produced by the generating units at a power generating station, calculated by subtracting the power consumed at the station for its own use (station use) from the gross generation.

Nitrogen oxides (NOx): Oxides of nitrogen that are a major component of air pollution and can be generated by the combustion of fossil fuels. They are also commonly referred to as nitrogen oxides.

Nonspecular conductors: Conductors that are specially treated to reduce light reflection. This treatment involves processes such as chemical treatment or surface modification to minimize shine and visibility.

Open-loop biomass: Biomass derived from agricultural livestock waste or solid, nonhazardous waste materials derived from cellulosic or lignin sources, including byproducts from wood or paper mill operations. This may also include lignin found in pulping liquors.

Organic Rankine cycle (ORC): Similar to the Rankine steam cycle, ORC uses thermal power to convert a working fluid to steam, which expands through a turbine in order to generate power. Different from Rankine steam cycle, ORC uses an organic fluid such as refrigerants and hydrocarbons instead of water.

Outage: The period during which a generating unit, transmission line, or other facility is out of service and not available for operation.

Particulate matter (PM): Tiny unburned fuel particles that form smoke or soot and can adhere to lung tissue when inhaled. They are a significant component of exhaust emissions from heavy-duty diesel engines.

Peaking capacity: The capacity of generating equipment that is typically reserved for operation during periods of highest daily, weekly, or seasonal loads. Some generating equipment may be operated as peaking capacity during specific times while serving loads on a continuous basis at other times.

Permeability: The measure of a porous medium's ability to transmit a liquid under a hydraulic gradient. In hydrology, it refers to the capacity of rock, soil, or sediment to allow the passage of water.

Petawatt (PW): A unit of power equal to one quadrillion (10^{15}) watts.

Petawatt hour (PW): A unit of energy equal to one quadrillion watt-hours.

Pollution: The introduction or presence of substances, contaminants, or pollutants into the environment, either directly or indirectly, resulting in adverse effects on ecosystems, human health, or the quality of air, water, or land. Pollution can occur from various sources, including industrial emissions, vehicle exhaust, improper waste disposal, chemical spills, and agricultural activities. It poses significant environmental challenges and requires effective mitigation measures and regulatory frameworks to minimize its impact and preserve the well-being of ecosystems and human populations.

Potential energy: The energy that an object possesses by virtue of its position or state. It is stored energy that can be converted into other forms of energy, such as kinetic energy, when the object undergoes a change in position or is acted upon by external forces.

Power block: A unit or module within a power plant that generates power. It typically consists of a set of interconnected components, such as a turbine, generator, and associated systems, which work together to convert a source of energy (such as solar, wind, water, biomass, geothermal, nuclear, or fossil fuels) into electrical power.

Power generation: (a) The process of converting other forms of energy into electrical power; (b) The quantity of electrical power produced, measured in kilowatt-hours (kWh).

Power plant: A facility designed to convert one or more forms of energy, such as electromagnetic, mechanical, chemical, and/or nuclear energy, into electrical power. Power plants typically consist of power generating systems, such as solar photovoltaic modules, (hydraulic, steam, gas) turbines, electric generators, reciprocating internal combustion engines, or other types of power generators and auxiliary equipment for converting sunlight, mechanical, chemical, and/or nuclear energy into electric energy, along with auxiliary equipment that supports the power generation process.

Power purchase agreement: A contractual agreement between a power plant and a large customer outlining the terms and conditions for the purchase and sale of electricity generated by the power plant.

Price: The amount of money or consideration-in-kind that is charged, paid, or offered in exchange for a product, service, or resource. It represents the value assigned to the item being exchanged and is determined by a range of factors, including supply and demand, production costs, market conditions, and competition.

Prime mover: A device or mechanism responsible for initiating or providing the primary source of power or motion in a system. It is the component that converts energy from one form to another, typically from a nonmechanical form (such as chemical, electrical, or thermal) into mechanical work. Examples of prime movers include engines, motors, turbines, or any device that generates mechanical energy.

Production well: A well drilled through a geothermal resource that is specifically designed and used to extract or produce geothermal fluids, such as hot water or steam, from the subsurface. These wells are crucial for harnessing geothermal energy for power generation or other direct uses.

Profit: The financial gain or benefit obtained from a business or investment after deducting all expenses, costs, and taxes. It represents the positive difference between the revenue generated and the total costs incurred, indicating the net income or earnings.

Rate of return: The annual rate of profit or gain on an investment, expressed as a percentage of the total investment amount. It measures the profitability and efficiency of an investment over a specific period and provides a basis for comparing the returns of different investment options.

Reconnaissance: A methodical process of gathering data, often associated with surface surveys, in which archaeological remains or natural resources are systematically identified, examined and plotted on a map. It involves preliminary investigations and observations to gain an initial understanding or overview of a particular area or subject.

Regulation: The governmental function of controlling or directing economic entities and activities through the process of rulemaking, enforcement, and adjudication. Regulations are established to ensure compliance with laws, promote public welfare, protect consumer rights, and maintain fair competition in various industries and sectors.

Reliability: The essential characteristic of an electric system that ensures a consistent and dependable supply of electricity to meet demand and energy requirements. It encompasses both adequacy, which refers to the ability to meet the expected load and demand, and security, which involves the ability to withstand and recover from unexpected disturbances or outages. Reliability is often measured by assessing the frequency, duration, and magnitude of any adverse impacts on consumer services.

Renewable energy: Energy derived from naturally replenishing sources, such as sunlight, wind, water (hydro), biomass, or geothermal heat, which are considered sustainable and have a lower environmental impact compared to fossil fuel-based energy sources.

Reservoir: An underground geological formation that contains a significant amount of fluid, such as water, oil, or natural gas, and can be tapped for extraction or storage purposes.

Retrofit: The process of upgrading or modifying an existing power plant, building, or equipment to improve energy efficiency, environmental performance, or comply with new regulations or standards.

Revegetation: Regrowing native plants, mainly trees and shrubs, by active restoration, natural process restoration, or both.

Revenue: The total amount of money received from various sources, including income from providing products and/or services, gains from the transactions or exchange of assets, interest, and dividends earned on investments, and other sources of income or inflows. It represents the overall income generated by entities, such as businesses, governments (including tax revenues), organizations, or individuals, regardless of the specific activities or sources involved.

Round-trip efficiency: The ratio, expressed as a percentage, of the useful energy output to the energy input into a process of energy production, conversion, or storage. The higher the round-trip efficiency is, the less energy is lost in the process.

Royalty: A payment made for the use or exploitation of property, such as intellectual property, copyrighted work, franchise rights, or natural resources. It is usually a percentage of the revenues or profits generated from the use of the property, and it compensates the owner or rights holder for granting permission or granting access to the property.

Run-of-river hydropower: A type of hydroelectric power generation that harnesses the natural flow and elevation drop of a river or stream to generate electricity without the need for large-scale reservoirs or dams.

Sales: The total amount of kilowatt-hours or electrical energy sold within a specific period of time. It is often categorized or grouped by classes of service, such as residential, commercial, industrial, and other categories. Sales may also include Public Street and highway lighting, sales to public authorities and railways, and interdepartmental sales within the electricity sector.

Slim hole: Small diameter wells drilled during the exploration phase to verify the existence of a productive geothermal resource and provide information about the geologic structure of the site. Slim hole wells are specifically designed to be less expensive and quicker to drill compared to full-diameter production wells. They serve as an initial assessment tool to determine the feasibility of geothermal development and provide valuable data for further decision-making and planning.

Smart grid: An advanced electricity distribution system that utilizes digital communication and automation technologies to monitor, control, and optimize the flow of electricity, improve grid reliability, enable demand response, and integrate renewable energy sources.

Socioeconomics: Research into the effects of both social and economic factors on individuals and communities. Socioeconomics begins with the assumption that economics is not a self-contained system but is embedded in society, polity, and culture.

Solar energy: Energy derived from the sun's radiation, which can be harnessed through various technologies such as photovoltaic (PV) systems that convert sunlight directly into electricity or solar thermal systems that utilize sunlight to generate solar heat for power generation, space and water heating, and other solar thermal applications.

Stability: The characteristic of a system or element whereby its output reaches a steady state over time. It represents the ability of a system, such as a power system, to withstand disturbances and return to a state of equilibrium. In the context of power systems, stability also relates to the capacity to generate restoring forces that are equal to or greater than the disturbing forces, ensuring that the system maintains a balanced state.

Standby service: Support service that is available, as needed, to supplement a consumer, a utility system, or another utility if a schedule or an agreement authorizes the transaction. The service is not regularly used.

Subsidence: A sinking of an area of the Earth's crust due to fluid withdrawal and pressure decline.

Substation: A facility in a power system where voltage levels are stepped up or down, and power is redistributed and controlled for efficient transmission and distribution to end-users.

Sulfur oxides: Compounds containing sulfur and oxygen, such as sulfur dioxide (SO_2) and sulfur trioxide (SO_3).

Sustainability: Economic development that takes full account of the environmental consequences of economic activity and is based on the use of resources that can be replaced or renewed and therefore are not depleted.

System (electric): Physically connected generation, transmission, and distribution facilities operated as an integrated unit under one central management or operating supervision.

Temperature gradient hole: A temperature gradient hole is a relatively slim and shallow hole (15–180 m deep) that aims to estimate the rate of increase in ground temperature with depth. The Geysers ("The" of "The Geysers" is always capitalized): A large geothermal steam field located north of San Francisco.

Terawatt (TW): A unit of power equal to one trillion (10^{12}) watts.

Terawatt hour (TWh): A unit of energy equal to one trillion watts hours.

Ternary pumped hydropower storage (T-PHS): An advanced pumped storage system that has a separate pump and a separate turbine on the same shaft and operates the pump and the turbine in the same direction. T-PHS makes transitioning between the generating mode and the pumping mode much faster and more flexible than conventional PHS.

Tidal energy: Renewable energy obtained from the kinetic energy of tides, which is harnessed using tidal turbines or barrages to generate power.

Transformer: An electrical device that transfers power between two or more circuits through electromagnetic induction, typically used to step up or step down voltage levels for efficient transmission and distribution of power.

Transmission grid: An interconnected infrastructure of power transmission lines and associated equipment for moving or transferring power in bulk over long distances from power generation to distribution substations or between different regions or countries for delivery over the distribution grid to power consumption or other electric systems.

Transmission: The movement and transfer of power across an interconnected infrastructure of lines and equipment, enabling power to be transported from power generation points to locations where it undergoes transformation before distribution to consumers or other electric systems.

Turbine: A rotary mechanical device that harnesses the energy of a fluid or air stream, such as water, steam, hot gas, or wind, to generate rotary mechanical power. Turbines operate on the principles of impulse, reaction, or a combination of both, converting the kinetic energy of the fluid or air into mechanical energy.

Utility demand: The amount of power or natural gas required by users at a specific point in time, typically expressed in kilowatts.

Voltage: The electrical potential difference between two points in an electrical circuit, measured in volts (V), which determines the intensity of electric current flow.

Watt: The electrical unit of power. The rate of energy transfer is equivalent to one ampere flowing under a pressure of 1 V at a unity power factor.

Watthour (Wh): A unit of measure for power consumption or generation. It represents the amount of power consumed from or generated by an electric circuit when a power of one watt is sustained for a duration of 1 h.

Wind power: Renewable energy obtained from the kinetic energy of wind, which is converted into electricity through wind turbines that capture the wind's force and rotate to generate electrical power.

Answer Key

Chapter 1. Solar PV Power

1e, 2c, 3c, 4b, 5e, 6d, 7a, 8d, 9b, 10c, 11a, 12c, 13b, 14c, 15b, 16c, 17b, 18d, 19e, 20b

Chapter 2. Solar Thermal Energy

1e, 2d, 3b, 4c, 5c, 6c, 7b, 8a, 9d, 10b, 11b, 12b, 13c, 14a, 15d, 16b, 17e, 18a, 19c, 20b, 21b, 22b, 23c.

Chapter 3. Wind Power

1c, 2b, 3c, 4c, 5c, 6b, 7b, 8c, 9b, 10a

Chapter 4. Hydropower

1c, 2a, 3c, 4b, 5b, 6c, 7c, 8e, 9b, 10c, 11d, 12b, 13b, 14c, 15b, 16d, 17e, 18b, 19c, 20b

Chapter 5. Bioenergy

1b, 2a, 3c, 4c, 5d, 6a, 7c, 8e, 9c, 10c, 11e, 12b.

Chapter 6. Geothermal Power and Heating

1c, 2d, 3e, 4b, 5c, 6d, 7c, 8b, 9d, 10d, 11e, 12d, 13e, 14a, 15c, 16d, 17c, 18d, 19c, 20a

Chapter 7. Energy Storage

1a, 2c, 3b, 4d, 5c, 6a, 7b, 8d, 9c, 10a, 11d, 12c, 13b, 14c, 15a, 16c, 17b, 18d, 19c, 20d.

P. Yang, *Renewable Energy*, https://doi.org/10.1007/978-3-031-49125-2

Index

P. Yang, *Renewable Energy*, https://doi.org/10.1007/978-3-031-49125-2